Lecture Notes in Physics

Springer-Verlag Berlin Heidelberg GmbH

Editorial Policy

The series *Lecture Notes in Physics* (LNP), founded in 1969, reports new developments in physics research and teaching -- quickly, informally but with a high quality. Manuscripts to be considered for publication are topical volumes consisting of a limited number of contributions, carefully edited and closely related to each other. Each contribution should contain at least partly original and previously unpublished material, be written in a clear, pedagogical style and aimed at a broader readership, especially graduate students and nonspecialist researchers wishing to familiarize themselves with the topic concerned. For this reason, traditional proceedings cannot be considered for this series though volumes to appear in this series are often based on material presented at conferences, workshops and schools (in exceptional cases the original papers and/or those not included in the printed book may be added on an accompanying CD ROM, together with the abstracts of posters and other material suitable for publication, e.g. large tables, colour pictures, program codes, etc.).

Acceptance

A project can only be accepted tentatively for publication, by both the editorial board and the publisher, following thorough examination of the material submitted. The book proposal sent to the publisher should consist at least of a preliminary table of contents outlining the structure of the book together with abstracts of all contributions to be included.
Final acceptance is issued by the series editor in charge, in consultation with the publisher, only after receiving the complete manuscript. Final acceptance, possibly requiring minor corrections, usually follows the tentative acceptance unless the final manuscript differs significantly from expectations (project outline). In particular, the series editors are entitled to reject individual contributions if they do not meet the high quality standards of this series. The final manuscript must be camera-ready, and should include both an informative introduction and a sufficiently detailed subject index.

Contractual Aspects

Publication in LNP is free of charge. There is no formal contract, no royalties are paid, and no bulk orders are required, although special discounts are offered in this case. The volume editors receive jointly 30 free copies for their personal use and are entitled, as are the contributing authors, to purchase Springer books at a reduced rate. The publisher secures the copyright for each volume. As a rule, no reprints of individual contributions can be supplied.

Manuscript Submission

The manuscript in its final and approved version must be submitted in camera-ready form. The corresponding electronic source files are also required for the production process, in particular the online version. Technical assistance in compiling the final manuscript can be provided by the publisher's production editor(s), especially with regard to the publisher's own Latex macro package which has been specially designed for this series.

Online Version/ LNP Homepage

LNP homepage (list of available titles, aims and scope, editorial contacts etc.):
http://www.springer.de/phys/books/lnpp/

LNP online (abstracts, full-texts, subscriptions etc.):
http://link.springer.de/series/lnpp/

M. Ristig K. Gernoth (Eds.)

Particle Scattering, X-Ray Diffraction, and Microstructure of Solids and Liquids

Springer

The presentations ranged over a spectrum of fields:
- Physics of quantum solids
- Theory of perfect crystals
- Computational physics
- Physics of liquid crystals
- Materials science
- Physical chemistry

The seminar provided a forum in which about fourty invited participants could evaluate the current status and development of the various branches of the application of scattering and diffraction. The setting of the meeting in the Physikzentrum Bad Honnef created a pleasant and productive atmosphere for fruitful discussions and exchange of experience and ideas. During the three days of the seminar twenty talks were presented by internationally recognized experts in their fields, coming from Europe and the United States. Six of the papers – extended to present more self-contained reviews – appear in this volume. The workshop included an evening talk on "Nikolaus von Cues and the Idea of Modern Natural Science," that offered a glimpse on the thinking of an extraordinary personality of more than half a millenium ago.

The meeting was made possible by the generosity of the WE-Heraeus-Stiftung. We express our sincere gratitude to Dr. Ernst Dreisigacker and Mrs. Jutta Lang from the WE-Heraeus-Stiftung for their excellent efforts in the general organization of the seminar. We also wish to extend our thanks to the staff of the Physikzentrum and to all speakers and participants for their contributions to a pleasant and fruitful workshop.

We wish to dedicate this volume to honor our colleague Wolfram Prandl, who participated and lectured on "Diffraction, Spectroscopy, and Thermodynamics" two months before his untimely death.

Koeln and Manchester, *Manfred L. Ristig*
July 2002 *Klaus A. Gernoth*

Contents

List of Contributors

Dr. Khaled AbuQuasem
Universität zu Köln,
Institut für Theoretische Physik,
50937 Köln, Germany
kq@thp.uni-koeln.de

Prof. Michael P. Allen
University of Bristol,
H. H. Wills Physics Laboratory,
Royal Fort, Tyndall Avenue,
Bristol BCS8 1TL, UK
m.p.allen@bristol.ac.uk

Aranzazu Maira-Vidal
C.S.I.C. / Instituto de Estructura de la Materia,
Dept. de Fisica Nuclear y Estadistica,
c/Serrano 123,
28006 Madrid, Spain
amaira@foton0.iem.csic.es

Prof. Dr. Ladislav Bohatý
Universität zu Köln,
Institut für Kristallographie,
50674 Köln, Germany
ladislav.bohaty@uni-koeln.de

Prof. Dr. Hans Boysen
Ludwig-Maximilians-Universität München,
Institut für Kristallographie,
Am Coulombwall 1,
85748 Garching, Germany
boysen@lrz.uni-muenchen.de

Prof. Dr. Eberhard Burkel
Universität Rostock,
Fachbereich Physik, Neue Materialien,
August-Bebel-Str. 55,
18051 Rostock, Germany
burkel@physik1.uni-rostock.de

Prof. John W. Clark
Washington University,
Department of Physics,
St. Louis, MO 63130, USA
jwc@wuphys.wustl.edu

Prof. Dr. Siegfried Dietrich
Max-Planck-Institut für Metallforschung,
Heisenbergstr. 1,
70569 Stuttgart, Germany
dietrich@mf.mpg.de

Dr. Ernst Dreisigacker
Wilhelm- und Else-Heraeus-Stiftung,
Postfach 1553,
63405 Hanau, Germany
dreisigacker@we-hereaus-stiftung.de

Dipl.-Phys. Joachim Dzubiella
Universität Düsseldorf,
Institut für Theoretische Physik II,
Universitätsstr. 1,
40225 Düsseldorf, Germany
joachim@thphy.uni-duesseldorf.de

Prof. Dr. Götz Eckold
Universität Göttingen,
Institut für Physikalische Chemie,
Tammanstr. 6,
37077 Göttingen, Germany
geckold@gwdg.de

Dipl.-Phys. Kalid Elisbihani
Universität Göttingen,
Institut für Physikalische Chemie,
Tammanstr. 6,
37077 Göttingen, Germany
kelisbi@gwdg.de

Dr. Damian J. J. Farnell
Ludwig-Maximilians-Universität
München,
Institut für Theoretische Physik,
Theresienstr. 37/399,
80333 München, Germany
farnell@theorie.physik.uni-muenchen.de

Dr. Ricardo Fernandez-Perea
C.S.I.C. / Instituto de Estructura de
la Materia,
Dept. de Fisica Nuclear y Estadistica,
c/Serrano 123,
28006 Madrid, Spain
ricardo@langran.iem.csic.es

Prof. Paolo Fornasini
Università di Trento,
Dipartimento di Fisica,
Via Sommarive 14,
38050 Povo (Trento), Italy
fornasin@science.unitn.it

Prof. Dr. Friedrich Frey
Ludwig-Maximilians-Universität
München,
Institut für Kristallographie,
Theresienstr. 41,
80333 München, Germany
frey@yoda.kri.physik.uni-muenchen.de

Prof. Dr. Werner Freyland
Universität Karlsruhe,
Institut für Physikalische Chemie,
Kaiserstr. 12,
76131 Karlsruhe, Germany
Werner.Freyland@chemie.uni-karlsruhe.de

Prof. Dr. Jürgen Fritsch
Universität Regensburg,
Institut für Theoretische Physik,
Universitätsstr. 31,
93040 Regensburg, Germany
juergen.fritsch@physik.uni-regensburg.de

Dr. Klaus A. Gernoth
UMIST, Department of Physics,
P.O. Box 88,
Manchester M60 1QD, UK
k.a.gernoth@umist.ac.uk

Dipl.-Phys. Holger M. Harreis
Universität Düsseldorf,
Institut für Theoretische Physik II,
Universitätsstr. 1,
40225 Düsseldorf, Germany
harreis@thphy.uni-duesseldorf.de

Dr. Sven E. Krüger
Universität Magdeburg,
Institut für Theoretische Physik,
Postfach 4120,
39016 Magdeburg, Germany
sven.krueger@physik.uni-magdeburg.de

Priv.-Doz. Dr. Karl Kürten
Universität Wien,
Institut für Experimentalphysik,
Boltzmanngasse 5,
1090 Wien, Austria
kuerten@ap.univie.ac.at

Priv.-Doz. Dr. Christos N. Likos
Universität Düsseldorf,
Institut für Theoretische Physik II,
Universitätsstr. 1,
40225 Düsseldorf, Germany
likos@thphy.uni-duesseldorf.de

Prof. Stephen W. Lovesey
Rutherford Appleton Laboratory,
ISIS Facility, Chilton,
Oxfordshire OX11 OQX, UK
s.w.lovesey@rl.ac.uk

Dr. Martin Müser
Universität Mainz,
Institut für Theoretische Physik,
Staudinger Weg 7,
55099 Mainz, Germany
martin.mueser@uni-mainz.de

Dr. Francesco Pederiva
Università di Trento,
Dipartimento di Fisica,
Via Sommarive 14,
38050 Povo (Trento), Italy
pederiva@science.unitn.it

Priv.-Doz. Dr. Wolf-Christian Pilgrim
Universität Marburg,
Institut für Physikalische Chemie,
Hans-Meerwein-Strasse,
35032 Marburg, Germany
pilgrim@mailer.uni-marburg.de

Prof. Dr. Wolfram Prandl †
Universität Tübingen,
Institut für Kristallographie,
Auf der Morgenstelle 10,
72076 Tübingen, Germany
wolfram.prandl@uni-tuebingen.de

Prof. Dr. Johannes Richter
Universität Magdeburg,
Institut für Theoretische Physik,
Postfach 4120,
39016 Magdeburg, Germany
johannes.richter@physik.uni-magdeburg.de

Prof. Dr. Manfred L. Ristig
Universität zu Köln,
Institut für Theoretische Physik,
50937 Köln, Germany
ristig@thp.uni-koeln.de

Prof. Dr. Gert Röpke
Universität Rostock,
Fachbereich Physik,
Universitätsplatz 1,
18051 Rostock, Germany
roepke@darss.mpg.uni-rostock.de

Prof. Dr. Hans-Eckhardt Schäfer
Universität Stuttgart,
Institut für Theoretische und Angewandte Physik,
Pfaffenwaldring 57,
70569 Stuttgart, Germany
schaefer@itap.physik.uni-stuttgart.de

Dr. Petra Schiebel
Universität Tübingen,
Institut für Kristallographie,
Auf der Morgenstelle 10,
72076 Tübingen, Germany
petra.schiebel@uni-tuebingen.de

Dipl.-Phys. Karin Schmalzl
Universität Regensburg,
Institut für Theoretische Physik,
Universitätsstr. 31,
93040 Regensburg, Germany
karin.schmalzl@physik.uni-regensburg.de

Prof. Dr. Winfried Schülke
Universität Dortmund,
Experimentelle Physik I,
Otto-Hahn-Strasse,
44221 Dortmund, Germany
schuelke@physik.uni-dortmund.de

Dr. Hans G. Senger
Universität zu Köln,
Thomas-Institut,
Universitätsstr. 22,
50932 Köln, Germany
CuSeng@uni-koeln.de

Prof. Ralph O. Simmons
University of Illinois,
Physics Department,
1110 West Green Street,
Urbana, IL 61801, USA
ros@uiuc.edu

Prof. Dr. Dieter Strauch
Universität Regensburg,
Institut für Theoretische Physik,
Universitätsstr. 31,
93040 Regensburg, Germany
dieter.strauch@physik.uni-regensburg.de

Dr. Yoshikazu Tanaka
RIKEN, Institute for Physical and Chemical Research,
1-1-1, Kouto, Mikazuki, Sayo, Hyogo,
629-5158, Japan
ytanaka@postman.riken.go.jp

Dr. David N. Timms
University of Portsmouth,
Physics Department,
St. Michael's Building,
Portsmouth PO1 2DT, UK
david.timms@port.ac.uk

Prof. Dr. Hans-Jürgen Weber
Universität Dortmund,
Institut für Physik,
Otto-Hahn-Str. 4,
44227 Dortmund, Germany
weberhj@fkp.physik.uni-dortmund.de

Prof. Dr. Roland Winter
Universität Dortmund,
Lehrstuhl für Physikalische Chemie,
Otto-Hahn-Str. 6,
44227 Dortmund, Germany
winter@steak.chemie.uni-dortmund.de

Prof. Dr. Helmuth Zimmermann
Universität Erlangen-Nürnberg,
Institut für Angewandte Physik,
Bismarckstr. 10,
91054 Erlangen, Germany
helmuth.zimmermann@krist.uni-erlangen.de

Scattering from Condensed Matter: A Brief Introduction

Klaus A. Gernoth[1] and Manfred L. Ristig[2]

[1] Department of Physics, UMIST,
P.O. Box 88, Manchester M60 1QD, United Kingdom
[2] Institut für Theoretische Physik, Universität zu Köln,
D–50937 Köln, Germany

Particle scattering and X-ray diffraction are universal tools in many branches of natural sciences for exploring and qualitatively and quantitatively analyzing the structure of matter. These physical mechanisms provide indispensible methods in the investigation of the physics of condensed matter, quantum fluids and solids, crystalline materials, classical liquids, perfect and imperfect crystals, liquid and solid surfaces, soft matter, such as polymers, etc. They are of equal importance in other fields of scientific research, in nuclear and medium-energy physics, physical chemistry, biophysics, chemistry, crystallography, microbiology and others. Scattering of particles, in particular of neutrons and electrons, and synchrotron radiation or X-ray diffraction permit not only the experimental exploration of the structure and properties of matter but make possible decisive tests of phenomenological models of microscopic mechanisms and of theoretical results derived by ab-initio many-body theories.

Large international research facilities, such as the Institute-Laue-Langevin, Grenoble, the Rutherford Appleton Laboratory, or DESY, Hamburg, provide the technical means to perform scattering and diffraction experiments. X-ray diffraction analysis is done by innumerable scientific institutes and research groups. Abundant literature is available that describes in detail the basic physics of particle scattering and diffraction, the measurement techniques, and the wide areas of applications. As background information for the beginner or the experienced specialist who wants to see the content of these lecture notes in a broader context we refer to some standard text books such as "Theory of Neutron Scattering from Condensed Matter" by S. W. Lovesey [1] and others [2–5]. Ref. [6] provides an elementary introduction to the general topic.

The last contribution in this volume may be read as a specialized but more advanced overview of typical features of inelastic neutron scattering experiments and selected recent applications to a very wide range of phenomena in differing scientific topics. This contribution focuses in particular on i) the exploration of structural excitations such as vibrations and dispersion laws of phonons in high-temperature superconductors and their role in soft-mode phase transitions, ii) diffusion processes characterized by single-particle motion of, for example, water molecules or organic molecules, iii) molecular rotations and tunneling exemplified by the properties of several solid phases of methane and the associated splitting of energy levels, and iv) polymer dynamics of poly-ethylene.

Contributions 2–6 of this volume present interesting new specific results of current theoretical and experimental work in various fields at the frontier of particle scattering and X-ray diffraction research.

Contribution 2 reports recent progress in neutron scattering and synchrotron X-ray scattering experiments on quantum crystals. The results reveal interesting new features of atomic dynamics and defects in solid helium isotopes. Presented are experimental and theoretical results on kinetic energies and excitation spectra of ^{3}He and ^{4}He quantum crystals and – for the first time – accurate Debye-Waller factors. This contribution also reports and discusses new and rewarding investigations on equilibrium point defects and their properties in these extraordinary quantum crystals. The contribution emphasizes the very fruitful interplay between the experimental findings and the theoretical predictions and shortcomings.

A generalized microscopic theory of perfect crystals based on first principles is outlined in contribution 3. The very promising algorithm focuses on the reduced one- and two-body density matrix elements of the scattering material as the central ingredients for a complete structural analysis of the correlated many-body system. It heavily exploits the crystallographic point and space group symmetries and their breaking in structural phase transitions. One of the most significant conclusions from these studies is that the full two-body density of a crystalline material is accessible experimentally by means of diffuse background scattering of a double-beam of neutrons or X-rays with appropriate wave vectors [7–16]. Present-day experiments on crystals are usually performed only with single-beams. They yield information on the form factors, i.e., on one-body quantities, but are insufficient for a determination of the reduced two-body density matrix elements which embody explicitly the correlation structure of the material. We note that the vast majority of symmetry-breaking irreducible order parameters does not appear in the one-body density but only in the two-body density for a system that undergoes a liquid-solid phase transition [17].

Contribution 4 presents a detailed theoretical study of high-temperature and high-pressure modifications of quartz. The study reflects the current state of the art of molecular dynamics simulations (for background literature see Refs. [18,19]). The results demonstrate that i) the instantaneous microscopic structure of a crystal (quartz) can deviate from the Gaussian average structure, even if domain disorder and/or crystal defects are absent, and ii) the definition of elastic constants may be not unique, if the crystal is subjected to large external pressures.

Molecular-dynamics simulations and Monte Carlo calculations [20] have been carried out by the author of contribution 5 for simple models of liquid crystals [21]. The review studies a variety of inhomogeneous systems such as planar interfaces with containing walls, equilibrium nematic-isotropic interfaces, and disclination defects in colloidal suspensions. The numerical results are compared with those of corresponding semi-analytical model theories based on ideas by Landau, de Gennes, and Onsager. Such models are still in their infancies and the inherent approximations and assumptions may be checked by stochastic calculations as performed in this contribution. The best, of course, would be to perform accurate ab-initio calculations for a given concrete system. However, liquid-crystal properties and the phase behavior of such systems are extremely sensitive to details of molecular structure and interactions. For this reason ab-initio calculations need extremely large sample sizes which require computer power that is beyond present-day computer capabilities. Adopting instead coarse-grained models is no true alternative, since the connection between the properties of such models and of the actual liquid-crystal system is blurred or lost. There is still a long way to go before one can attempt a quantitative comparison between experimental and theoretical results.

Disorder effects in crystals and quasicrystals are described and discussed in contribution 6. Experiments in this field of research are usually performed by employing diffuse diffraction methods [22]. A large variety of structural fluctuations exists that can cause departures from perfectly ordered arrays of structurally identical units in real space (crystals) or in higher-dimensional hyperspace (quasicrystals). One distinguishes different classes of disorder, such as i) chemical, compositional, and substitutional replacements (examples are alloys and mixed crystals), ii) positional displacements (by strains, modulations, phonons, for example), iii) domain structure (more generally lattice disorder), and iv) disorder due to substructures. The contribution explains these effects and their applications in detail.

Due to the limited seminar time no lecture was presented on scattering problems in nuclear and medium-energy physics despite their close analogy to scattering and diffraction problems covered in this volume. A typical example from this very active field of research that demonstrates the close kinship is the quasi-elastic electron-nucleon scattering from carbon at different momentum transfers [23]. Such ^{12}C experiments yield valuable information on the nuclear momentum distribution (and the off-diagonal elements of the one-body density matrix). The measured cross sections are analyzed within the impulse approximation [24] (ignoring final-state effects) and are preferentially described by a so-called Y-scaling variable [25]. However, the data analysis becomes complicated by excitations of subnucleonic degrees of freedom at medium energies, since they lead to a breakdown of the Y-scaling properties. Such scale-breaking occurs also in condensed matter systems with atomic or molecular constituents which have both translational and internal degrees of freedom. An excellent example is furnished by liquid molecular hydrogen [24,26].

In contrast to molecular-hydrogen crystals at low temperatures, which have been extensively studied by scattering techniques [27], the liquid H_2 phase has long resisted to reveal its detailed microscopic structure [28]. The experimental complications arise – as in the nuclear case of ^{12}C – from the presence of intra-molecular vibrations and rotations causing scale-breaking. Only recently successful neutron scattering measurements have been designed and performed, where the molecular excitations are sufficiently suppressed [28,29]. A detailed analysis of the cross section data now permits a quantitative determination of the liquid structure function of molecular hydrogen close to the triple point. Results of a first quantitative examination of this correlation function and a comparison with theoretical predictions are reported in Refs. [28,30].

The molecular hydrogen problem demonstrates that well established particle scattering methods are not only powerful tools for a structural analysis of physical and biological materials of ever increasing complexity but can still be employed and adopted for investigations of problems of fundamental interest in quantum many-body physics.

References

1. S. W. Lovesey, "Theory of Neutron Scattering from Condensed Matter," Vol. 1, Clarendon Press, Oxford, 1984.
2. G. L. Squires, "Introduction to the Theory of Thermal Neutron Scattering," Dover Publications, Mineola, New York, 2000.
3. J. M. Cowley, "Diffraction Physics," North-Holland, Amsterdam, 1981.
4. E. Burkel, *Rep. Prog. Phys.* **63** (2000), 171.
5. E. Burkel, *in* "Springer Tracts in Modern Physics," Vol. 125 (Springer-Verlag, Berlin, 1991).
6. 27. IFF-Ferienkurs "Streumethoden zur Untersuchung kondensierter Materie" (Forschungszentrum Jülich, 1996).
7. W. Bauspiess, U. Bonse, W. Graeff, and H. Rauch, *J. Appl. Cryst.* **10** (1977), 338.
8. U. Bonse, *in* "Neutron Interferometry" (U. Bonse and H. Rauch, Eds.), p. 3, Clarendon Press, Oxford, 1979.
9. H. Rauch, *in* "Neutron Interferometry" (U. Bonse and H. Rauch, Eds.), p. 161, Clarendon Press, Oxford, 1979.
10. W. Schülke, U. Bonse, and S. Mourikis, *Phys. Rev. Lett.* **47** (1981), 1209.
11. W. Schülke, *Solid State Commun.* **43** (1982), 863; Erratum: *Solid State Commun.* **44** (1982), 1130.
12. G. Materlik, and J. Zegenhagen, *Phys. Lett. A* **104** (1984), 47.
13. H. Spalt, A. Zounek, B. N. Dev, and G. Materlik, *Phys. Rev. Lett.* **60** (1988), 1868.
14. U. Bonse, *Physica B* **151** (1988), 7.
15. W. Schülke and A. Kaprolat, *Phys. Rev. Lett.* **67** (1991), 879.
16. H. Rauch and S. A. Werner, "Neutron Interferometry: Lessons in Experimental Quantum Mechanics", Clarendon Press, Oxford, 2000.
17. K. A. Gernoth, *Ann. Phys. (N.Y.)* **285** (2000), 61; *Ann. Phys. (N.Y.)* **291** (2001), 202.
18. D. Frenkel and B. Smit, "Understanding Molecular Simulation: From Algorithm to Applications, Academic Press, San Diego, 1996.
19. M. P. Allen and D. J. Tildesley, "Computer Simulation of Liquids," Clarendon Press, Oxford, 1989.
20. D. P. Landau and K. Binder, "A Guide to Monte-Carlo Simulations in Statistical Physics," Cambridge University Press, Cambridge, 2000.
21. P. G. de Gennes and J. Prost, "The Physics of Liquid Crystals," Clarendon Press, Oxford, 1995.
22. L. H. Schwartz and J. B. Cohen, "Diffraction from Materials," Academic, New York, 1977.

23. D. B. Day et al., *Phys. Rev. Lett.* **59** (1987), 427.
24. R. N. Silver and P. E. Sokol, Eds., "Momentum Distributions," Plenum, New York, 1989.
25. G. B. West, *Phys. Rep.* **18C** (1975), 263.
26. W. Langel, D. L. Price, R. O. Simmons, and P. E. Sokol, *Phys. Rev. B* **38** (1988), 11275.
27. I. F. Silvera, *Rev. Mod. Phys.* **52** (1980), 393.
28. F. J. Bermejo, *Phys. Rev. Lett.* **84** (2000), 5359.
29. F. J. Bermejo et al., *Phys. Rev. B* **47** (1993), 15097; *Phys. Rev. B* **60** (1999), 15154.
30. M. L. Ristig, T. Lindenau, and L. Szybisz, *in* "Condensed Matter Theories," Vol. 17, Nova Science, Huntington, New York, in press.

Scattering Studies of Condensed Helium Isotopes

Ralph O. Simmons

Department of Physics and Frederick Seitz Materials Research Laboratory, University of Illinois at Urbana–Champaign, Urbana, IL 61801, USA

Abstract. Both neutron and synchrotron X–ray scattering reveal important features of atomic dynamics and defects in condensed He isotopes. Neutron work has obtained kinetic energies and detailed excitation spectra. Recently, X–ray work has produced accurate Debye–Waller factors of crystal ^{3}He and ^{4}He and has provided the first phonon excitation data on ^{3}He. X–ray diffraction from He isotope crystals also provides the only direct means to measure equilibrium point defect content, and therefore defect free energy. Understanding of all these phenomena is assisted by sophisticated Monte Carlo simulations. Thereby, future experiments on these interesting quantum systems are suggested, and reciprocally, motivations for future theoretical work are provided.

1 Introduction

In diverse parts of condensed matter physics, X–ray and neutron scattering studies of condensed He isotopes show opportunities for new understanding, through comparison with sophisticated computations of properties of these quantum systems. The neutron Compton profile, obtained from recoil spectra at eV incident neutron energies, can be analyzed to obtain directly the atomic momentum distribution, and hence the kinetic energy, $\langle E_k \rangle$ [1,2]. Precise results have already been obtained in fcc, hcp, and bcc He, as well as for normal He fluids [3,4]. Path–integral Monte Carlo and related simulations of $\langle E_k \rangle$ using modest sample sizes ($N \sim 500$) can explore extreme conditions of density and at the same time be guided by quantitative experimental results. Monte Carlo methods are discussed elsewhere in this book [5].

Synchrotron X–rays make possible several extensions of dynamical studies of He, for tiny crystals, for extreme ranges of scattered intensity, beyond the reach of usual triple–axis coherent inelastic neutron techniques [6]. Crystals of ^{3}He and ^{4}He are equally accessible. Over large length scales, precise systematic measurements of distributions of atoms about lattice sites, and hence mean–square atomic displacements, $\langle u^2 \rangle$, in hcp and fcc ^{3}He and ^{4}He crystals over a wide range of densities show systematic differences from PIMC computations using modest sample sizes [7,8]. These differences are removed through extrapolations to the thermodynamic limit [9]. Such extrapolations produce little change in $\langle E_k \rangle$, which is dominated by short–scale high–energy excitations in these crystals. However, at the limit, unusual temperature dependencies for $\langle u^2 \rangle$ have been reported in simulations, which would place existing precise measurements, at moderate density, in a crossover regime intermediate between quantum and classical. At the same time, they are consistent with an excitation model having phonon widths comparable to the phonon energies.

Coherent inelastic X–ray scattering from phonon excitations has been investigated in both hcp ^{3}He and ^{4}He crystals at a density within the range of $\langle u^2 \rangle$ studies. Results, including relative intensities, have been interpreted within the framework of a multi–phonon excitation analysis [10,11]. It is clear that existing self–consistent phonon models for crystal He are unsatisfactory, and give seriously wrong density dependence. Unexpected results were found for the [001] direction in hcp crystals: X–ray excitation widths differ in zones usually considered to be equivalent. Indeed, the fitted widths are comparable with the excitation energies. This is consistent with the extrapolated PIMC inferences already noted [9]. Evidently, these phenomena merit both further X–ray experiments at higher resolution and new computations to investigate multi–phonon excitations.

The properties of possible point defects in He crystals have long been studied analytically. Instructive but phenomenological models [12] and analogies to other physical systems [13] were pursued by pioneers. However, microscopic structural models for the defects have been relatively unexplored until recently [14,15]. Those models, as well as the earlier analytical studies, are primarily for low densities, where the defects are expected to be most unusual. On the experimental side, there have been many defect investigations, on both He isotopes and their mixtures. These have employed diverse techniques. However, for defects, the experimental situation remains controversial [16]. Claims of evidence for zero–point and/or highly non–localized defects have been made following inferences from ultrasonic and NMR measurements. Defect content inferred from analysis of these data and from calorimetric and thermal conductivity data have not appeared to be consistent with isochoric thermal pressure and in particular, with direct X–ray studies of net defect content. New isochoric thermal pressure work, and especially, X–ray diffraction work on close–packed ^{4}He at the borders of the wide molar volume range 10.4–20.8 cm^3 has been done [8,17]. From these data, one gets absolute net vacancy concentrations and thereby obtains the vacancy free volume of formation as a function of density over a broad range. New microscopic modelling is thereby encouraged of these unusual defect structures.

Considerations here are limited of liquid states and of the many techniques besides diffraction to study defects in He. Omitted are such fascinating phenomena as crystallization waves at the fluid–crystal interface [19], faceting in crystal growth [20], neutron diffraction from magnetic structures in crystal ^{3}He [21], and X-ray study of electronic excitations in condensed He [22]. Outside the scope of this chapter are scattering investigations of superfluid ^{4}He [23].

1.1 Noble Gas Solids

The noble gases are a family of weakly–interacting atoms which provide excellent opportunities for investigating the relative importance of many phenomena. The pair interaction potentials are well–determined and multi–body interactions

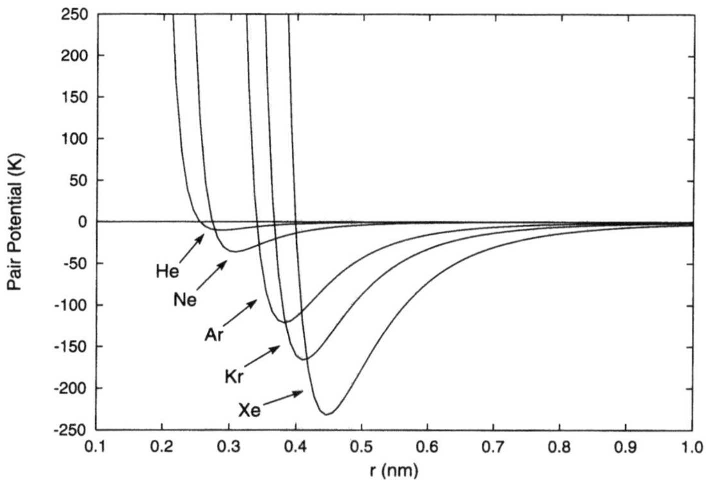

Fig. 1. Lennard-Jones models for pair interactions for the noble gases. The actual potentials are rather different [24]. In the condensed state, multi–body interactions become relatively more important as mass increases. He forms the most weakly–bound solid.

have been much investigated [24]. For computational simplicity, many older theoretical studies employed a Lennard–Jones (LJ) potential, which has only two parameters. One, ϵ, characterizes the depth of the pair potential well, and the other, σ, characterizes the interatomic distance at the minimum. Such a potential can illustrate qualitatively several points to be made here. In Fig. 1 are shown the LJ potentials for the family. As the mass increases, from He to Xe, the pair parameters ϵ and σ both increase in absolute value.

For He, the pair interaction is very weak and very shallow. The LJ ϵ is 10.22 K, while the depth of the actual potential is 10.956 K. The properties of the condensed states of He are notably affected. Consider that the thermal wavelength, λ_T, of a ^{4}He atom of mass m at 4 K is given by

$$\lambda_T = h/\sqrt{2\pi m k_{\rm B} T} \approx 0.5 \text{ nm} , \tag{1}$$

where h and $k_{\rm B}$ are the Planck and Boltzmann constants, respectively. This is a distance well beyond the minimum of the He potential shown in Fig. 1. Indeed, He does not form a solid, even at $T = 0\,$K, unless external pressure is applied to reduce the molar volume sufficiently. In that condition, the total energy of solid He has a considerable component of kinetic energy. This energy is volume–dependent, providing an internal quantum pressure [25].

To discuss such effects in the context of corresponding states, de Boer [26] defined a quantum parameter

$$\Lambda^* = \frac{h}{\sigma\sqrt{m\epsilon}} . \tag{2}$$

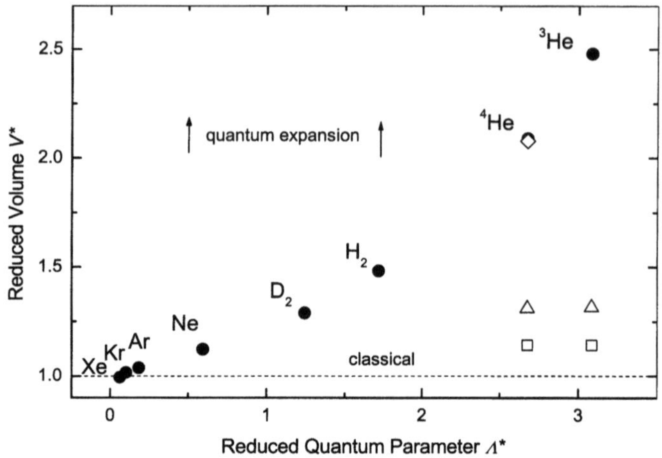

Fig. 2. Measured maximum molar volumes of solids plotted in reduced units V^* versus Λ^* The massive atoms form classical solids, while the least massive show quantum expansion [26] to the volumes indicated (•). Present work on He was at volumes marked by other symbols: Debye-Waller (□), phonons (△), kinetic energy $\langle E_k \rangle$ (◇).

Reduced parameters can also be formed, e.g. a reduced volume

$$V^* = \frac{V}{N_A \, \sigma^3} \, , \tag{3}$$

where N_A is the Avogadro number.

It is of interest to examine the relation observed in the condensed noble gases between the reduced volume V^* and the quantum parameter Λ^*. The results are shown in Fig. 2. For each element, the values of V^* were obtained from typical molar volumes observed in the condensed state, with the He isotopes represented at the condition of solidification at lowest sufficient pressure. It is seen that the massive members of the family (Xe, Kr, Ar) cluster around a classical volume, while the isotopes ^{4}He and especially ^{3}He are considerably expanded by the quantum pressure. Also shown are the molecular quantum solids H_2 and D_2 in their $J = 0$ states (central intermolecular interactions).

Most of the X–ray scattering data to be discussed here were collected on He crystals compressed to a volume less than maximum, one more equivalent to that of Ne, which shows only mild quantum expansion. Volumes for the He crystals discussed here are indicated by the respective symbols in Fig. 2. However, it does not follow that such compressed He crystals are as simple as Ne. This is illustrated in Fig. 3, which shows the results of PIMC simulations of spherically–averaged pair distribution functions $g(r)$ for Ne and for ^{4}He, respectively [27]. For natural Ne at 20.3 K one sees the usual $g(r)$ found for a fcc crystal. In particular, note the distinct second–neighbor peak (marked by the right-hand arrow). However, in the case of crystal ^{4}He, the functions $g(r)$ appear remarkably similar to those for liquid. Only upon compression of ^{4}He to V around 10 cm^3

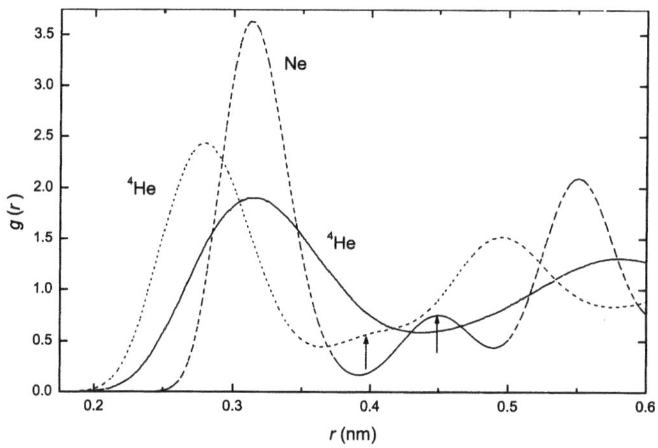

Fig. 3. Spherically–averaged pair distributions, $g(r)$, in crystals, from PIMC computations [27]. Ne at 20.3 K (*dotted*) shows behavior typical of monatomic fcc solids, including the obvious presence of a second–neighbor peak (*right-hand arrow*). On the other hand, crystal ^{4}He has fluid–like shape for $V = 15.06$ cm^3, 7 K (*solid line*), which only begins to appear crystalline (*left-hand arrow*) as it is compressed to higher density, $V = 10.04$ cm^3, 20 K (*dashed*). For $g(r)$ in bcc phase ^{3}He, see [28].

(the □ point on Fig. 2) does a hint of the second–neighbor peak appear (left-hand arrow). The success, for fcc Ne, of dynamical models such as forms of the self–consistent–phonon (SCP) model [29,30] does not therefore guarantee success also for compressed He crystals.

1.2 Motivations for Phonon and Defect Studies in Solid He

There are very active significant advances in fundamental theoretical methods which can treat phonons in quantum solids, the dynamics of liquids, and crystalline defect properties – all on a common basis. Examples of such computational methods are Path Integral Monte Carlo (PIMC) and variational Shadow Wave Functions (SWF) [31,32]. Such methods can be tested through critical comparisons over the widest possible range of experimental conditions, including the examination of isotopic differences. On the experimental side, samples of extremely high purity can be prepared for measurement.

The boson solid ^{4}He has a very wide range of densities and temperatures which can be reliably and precisely studied in the laboratory. Unlike ^{4}He, the fermion ^{3}He has in addition a broad range of bcc phase in its equilibrium phase diagram for density– and T–dependent studies. The ^{3}He bcc phase is the most quantum and liquid–like of 3D crystalline phases anywhere in nature [25,28]. Recall, from Fig. 3, that the ^{4}He fcc phase already shows some liquid character.

Deductions from calorimetry [33] suggest that Debye characteristic temperatures of ^{3}He and ^{4}He do not scale precisely with the mass ratio $\sqrt{4/3}$, even though the bare pair interaction for the two isotopes is identical. In a harmonic approximation such experimental results, obtained at low T, weight the phonon spectrum by ω^{-3}, where ω is a phonon frequency. Further, it cannot be confirmed either from ^{3}He X–ray Debye–Waller [7] or from X–ray phonon work [11] done so far, that hcp phonon frequencies for ^{3}He and ^{4}He necessarily scale exactly by $\sqrt{4/3}$. In a harmonic picture, one expects that the Debye–Waller effects weight the phonon spectrum by ω^{-1} and by ω^{-3} at low– and high T's, respectively [34].

It has been found that anharmonicities are seen through multi–phonon and interference effects in the excitation spectra obtained from inelastic scattering studies. Some aspects of those studies will be discussed below. There are important open questions.

Finally, beyond the properties of ideal homogeneous phases, there are thought to be some qualitative differences in the point defect properties of ^{3}He and of ^{4}He [12,14]. Beyond fermion and boson differences, these may be related to large contributions of atomic dynamics to defect states. When a common computational basis is available to treat these, and the experimental phenomena can be reproduced, confidence will be increased that the relevant factors governing behavior of these interesting systems are understood.

1.3 Theoretical Models

It is not within the scope of this chapter to describe the models themselves. Rather, the purpose is to describe selected phenomena, to compare them to existing models, and to point out some areas likely to reward further study.

Self–consistent Phonon Theories. In the 1960's came major developments in treating highly anharmonic lattice dynamics. First were new computational and perturbation methods which led to better microscopic understanding of the thermodynamics [35]. Then came formulations which came to be called self–consistent phonon theories. In fact, they had varying degrees of self–consistency. In solid Ne, they explained some explicit anharmonicities, for example, the T dependence of phonon energies at constant V [29,30,36]. Effective–potential work on Ne brought remarkable agreement with precise experiment at constant P [37]. In solid He, SCP computations achieved qualitative success with one–phonon dispersion, and they explained some excitation curve shapes as multi–phonon interference effects [23].

However, in various He solids 20% errors were common in one–phonon energies. Note, this is larger than raw isotope mass effect $\sqrt{4/3} = 1.15$. Such inaccuracy does not sensitively test changes in the effective potential with change of mass. While such errors are significant, they are less disturbing than the failure of existing SCP models to give realistic density dependencies for phonon energies. These matters will be considered below in Sect. 3.1.

Finally, from a comprehensive point of view, there are two serious shortcomings of such theories workable in perfect crystals. First, they provide no consistent link to properties of the liquid state. Second, they have no straightforward link to point– or line–defect properties.

Monte Carlo Methods, PIMC and SWF Variational. As more powerful computers became available, Monte Carlo methods could be developed and applied. For the simulation of condensed He phases, these have provided many interesting results [31]. They obtained accurate melting curves of both ^{3}He and ^{4}He over density ranges of a factor five. Note, this is an extremely sensitive test for very accurate values of total energy of both liquid and crystal phases. They have obtained kinetic energies, $\langle E_k \rangle$, to a few percent accuracy, in both solid and liquid phases [3,4] and for both isotopes [38]. They have treated superfluid ^{4}He and magnetic properties of ^{3}He. When extrapolated to the thermodynamic limit, they obtain mean–square–deviation values, $\langle u^2 \rangle$, accurately in agreement with measured Debye–Waller factors [9]. Finally, these methods are applicable not only to solid and liquid, but also to point defect calculations on a consistent basis [14]. In later sections, these advantages will be seen in the context of comparisons to experimental studies.

2 Coherent Scattering Studies

With only two electrons per atom, He is a weak scatterer of X–rays. Only with the advent of synchrotron sources have systematic studies of scattered intensities become practical. Integrated X–ray intensities I_{hkl} have been collected from a considerable number of He isotope crystals, by Venkataraman [8,18,39] and by Arms [7,40]. For the latter, corresponding (P,T) points, for samples at the given pressure, P, and T, are shown in Fig. 4, together with the phase boundaries for ^{4}He and ^{3}He, respectively. Both hcp and fcc phases were studied, with single crystals prepared in a Be sample cell shown in Fig. 5 [41,42]. A typical cell

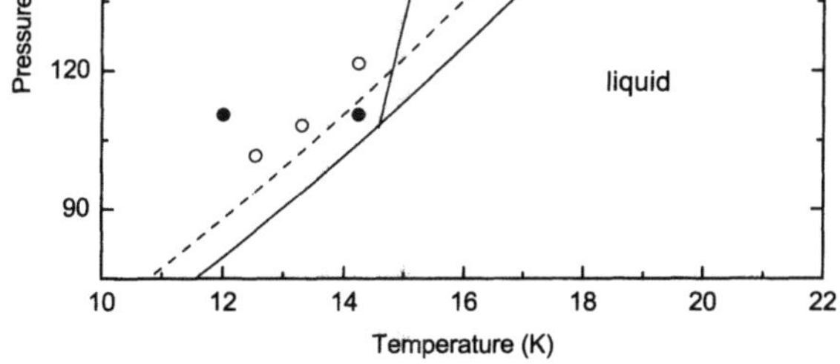

Fig. 4. Superimposed ^{3}He and ^{4}He phase diagrams show that the smaller mass isotope (*dashed line*) has the larger quantum internal pressure and therefore melts at a higher P for a given T. The points (P,T) denote conditions at which Debye–Waller X–ray data were collected by Arms [7] for ^{3}He (∘) and ^{4}He (•).

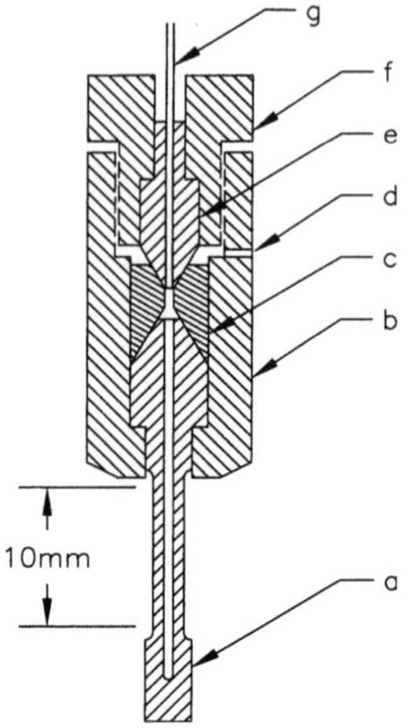

Fig. 5. Pressure cell [41] used for synchrotron X–ray studies of helium crystals. a) cell of sintered Be clamped to refrigerator at blind end, b) Maraging steel (MS) collar, c) stainless steel insert, d) vent, e) MS fitting for capillary fill line g, f) MS nut, g) line to external gas handling system. The cell is enclosed within thermal radiation shields and a vacuum jacket.

has 0.8 mm I.D. and 2.0 mm O.D., and can be used up to 300 MPa, which is sufficient to reach fcc phases of both He isotopes. It is mounted on an orientable closed–cycle refrigerator, in a way that permits a temperature gradient to be maintained during crystal growth, and precise temperatures maintained during measurements.

Attenuation of coherent (Bragg) scattering from a crystal arises owing to displacement of scattering units from their mean equilibrium positions. Such displacements arise from both zero–point and thermal excitation. The integrated intensity, I_{hkl}, of a Bragg peak (hkl) is reduced by the Debye–Waller factor,

$$I_{hkl} \propto \exp\left(-2M_Q\right) , \tag{4}$$

where $\boldsymbol{Q} = \boldsymbol{k_f} - \boldsymbol{k_i}$ is the wave vector transfer of the scattered probe having wave vector k. M_Q is the Debye–Waller factor along the direction of $\boldsymbol{Q}$. The Debye–Waller factor can be expanded schematically as

$$2M_Q = \langle {u_Q}^2 \rangle \, Q^2 \; - \; (1/12)(\langle {u_Q}^4 \rangle - 3\langle {u_Q}^2 \rangle^2)Q^4 \; + \; O(Q^6) \, , \tag{5}$$

where $\langle {u_Q}^2 \rangle$ is the (T– and P–dependent) mean square atomic deviation along the direction of $\boldsymbol{Q}$. If the distribution of deviations of the atoms from their equilibrium positions is Gaussian and isotropic, the Debye–Waller factor reduces to

$$2M_Q = \langle {u_Q}^2 \rangle \, Q^2 \, . \tag{6}$$

An empirical test for this for a sample at a given (P, T) is to plot the integrated intensities, appropriately corrected for any form factors, probe polarization effects, and geometrical factors, in the form

$$\ln I_{hkl} = \text{constant} - \langle {u_Q}^2 \rangle Q^2 \,. \tag{7}$$

If individual I_{hkl} values lie off a line in any systematic way, this may be an indication of anisotropy of the mean square deviations in the crystalline lattice. A random distribution of differences can be used to estimate uncertainty in the value of $\langle u^2 \rangle$.

One way to represent T dependence of $\langle u^2 \rangle$ at constant V in a sensitive manner is to calculate an equivalent Debye characteristic temperature, Θ_M, for each $\langle u^2 \rangle$ value, and then to plot the resulting Θ_M's versus T. Each $\Theta_M \equiv Tx$ is calculated from the expression

$$T \langle u^2 \rangle = \frac{3\hbar^2}{m k_{\mathrm{B}}} \left[\frac{1}{4x} + \frac{1}{x^3} \int_0^x \frac{\varepsilon \, \mathrm{d}\varepsilon}{\mathrm{e}^{\varepsilon} - 1} \right] \,. \tag{8}$$

When $\langle u^2 \rangle$ data on samples at different molar volumes are compared, it can be justified to apply a correction of conventional Grüneisen parameter type. Strictly, one should use the parameter applicable to the X–ray phonon spectrum, γ_M. In the case of harmonic lattice dynamics, this is well–defined in terms of frequency moments [34]. Such a procedure also works for the somewhat anharmonic crystal Ne. However, for highly anharmonic He it is necessary at present to be content with using the parameter applicable to heat capacity measurements, γ_{C_V}. This does have the advantages that precise data are available over wide ranges of temperature and molar volume for ^{4}He [43], and for smaller ranges for ^{3}He [44]. With extrapolations of the corresponding Θ_{C_V}'s to $T = 0$, the following fit to these data is obtained:

$$\gamma_{C_V} = 0.8114 + 0.09690\, V \qquad 26 \leq V(\mathrm{cm}^3) \leq 13.7 \,. \tag{9}$$

Applying this V–dependent parameter to the usual Grüneisen relation, the scaling between two different states 1 and 2 using a given phonon frequency associated with a given volume becomes

$$\ln(\omega_2/\omega_1) = 0.0969(V_1 - V_2) + 0.8114 \ln(V_1/V_2) \,. \tag{10}$$

2.1 Debye–Waller Measurements

Some He crystals are excellent, having narrow and symmetrical rocking curves. Figure 6 shows results on such an fcc ^{3}He sample [7], taken at the Brookhaven NSLS beamline X–14A [45]. Two data sets are shown. The rocking curve widths are indistinguishable from those taken on the same beamline with a Si sample.

Even when crystals are not excellent, careful measurements of the Bragg integrated intensities give satisfactory results. Figure 7 shows results on an hcp ^{3}He sample, taken at the Argonne APS beamline 1–BM [46]. A typical mosaic

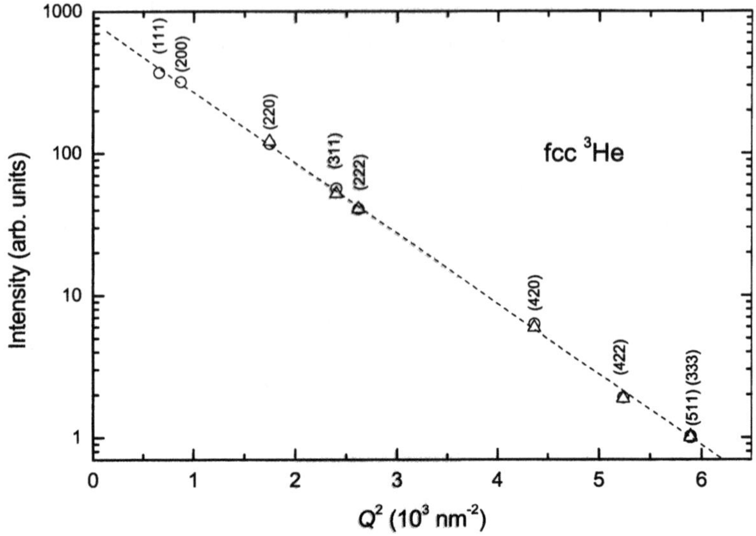

Fig. 6. Integrated X–ray intensity data for fcc ^{3}He of molar volume 11.578 cm^3 at 18.13 K. This crystal was of excellent quality. The value of $\langle u^2 \rangle$ is obtained from the negative slope of the line (7). Both data sets (Δ and $\circ$) yield the same value.

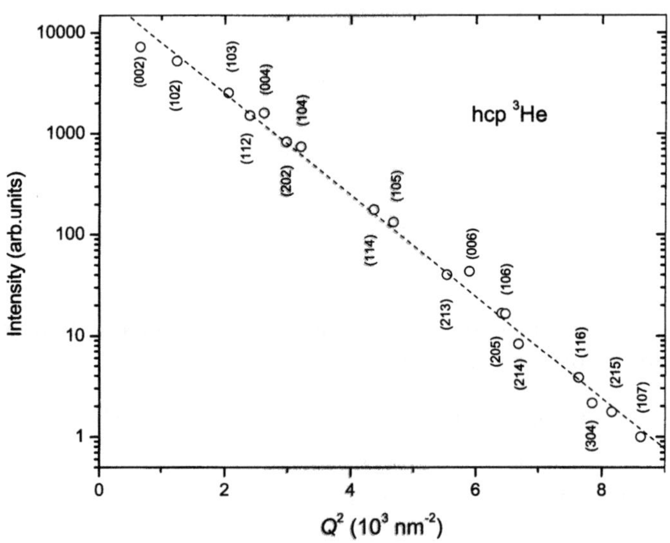

Fig. 7. Integrated X–ray intensity data for hcp ^{3}He of molar volume 11.607 cm^3 at 16.90 K. The slope of the line, and therefore the value of $\langle u^2 \rangle$, is well–determined, even though considerable crystal mosaic produced varying rocking curve profiles at different (hkl) settings of the crystal. The data cover a very large range in Q.

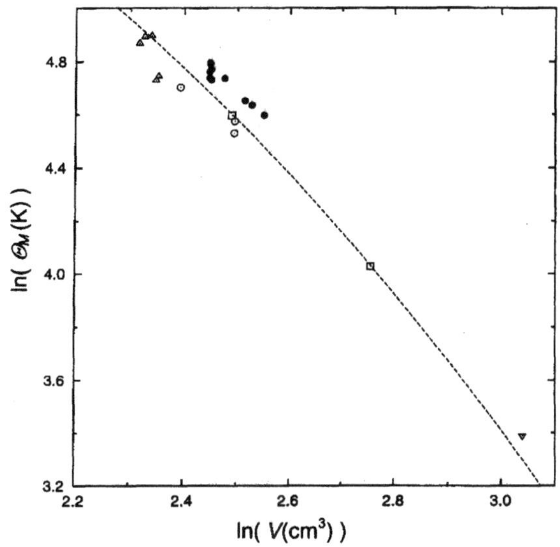

Fig. 8. Comparison of the average lattice dynamics of He isotopes through the Debye X–ray characteristic temperature, Θ_M, over a large range of V. X–ray data are ^{3}He data from [7](•) and ^{4}He data from [7](○), [8](△), and [47](▽). Neutron data on ^{4}He are from [48](□). The dashed line is of form (10) at constant T, from heat capacity work. Note, the data were taken at different T's, which contributes some scatter.

width for this crystal was 0.2 deg. Note, the data are plotted after necessarily large corrections were made for the Q–dependence of the X–ray form factor [7]; the original intensity data covered a dynamic range of 10^6. In such circumstances, severe demands are made upon the linearity of detectors, the precision of attenuation techniques, the suppression of background, and mechanical and thermal stability of the experimental arrangement, including stability while the sample cell is reoriented to gather data at different values of (hkl). The specific (hkl) values are labelled on Fig. 7; one sees that they correspond to considerable reorientations of the single crystal.

Measurements for both ^{3}He and ^{4}He crystals can be summarized empirically by use of the Debye characteristic temperature, $\Theta_M(V,T)$. For each individual measurement of $\langle u^2 \rangle$, the relation (8) was used to calculate a Θ_M. It turns out that for a given He isotope, the T dependence is less than the V dependence, so the data are presented first versus V, in Fig. 8. Besides synchrotron X–ray data [7,8,39,40,47], the single set of neutron data are shown [48]. The dashed curve is the relation generated by (10) with $k_B\Theta_M = \hbar\omega$ for a fixed T. It fits the general trend of the data reasonably well. Moreover, the values for ^{3}He lie above those for ^{4}He, although a ratio of $\sqrt{4/3}$ is neither confirmed nor denied. Note, some of the scatter in Fig. 8 arises for data taken at different T's.

Finally, Fig. 9 shows, for the same measurements shown in Fig. 8, the temperature dependence of Θ_M calculated using (8), after scaling to $V = 11.5$ cm^3. This V is symbol □ in Fig. 2. Here, the sensitivity of the calculation at low tem-

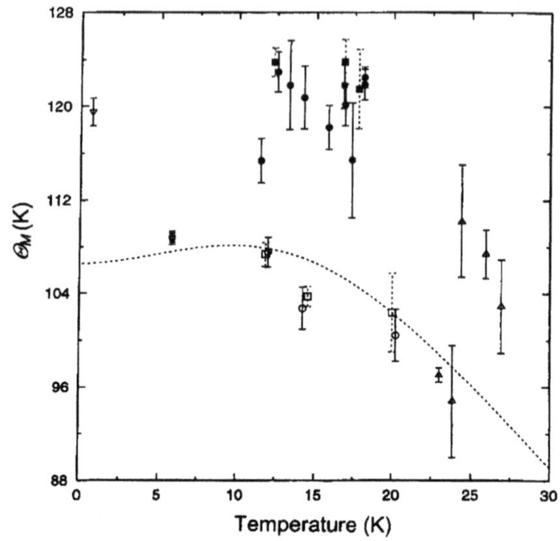

Fig. 9. Comparison of the average lattice dynamics of He isotopes through the Debye X–ray characteristic temperature, Θ_M. To display the T–dependence sensitively and to compare ^{3}He and ^{4}He, the data of Fig. 8 are scaled to a common $V = 11.5$ cm^3 using the Grüneisen parameter. Symbols are same as Fig. 8. Also shown (*dashed*) is a line corresponding to the PIMC T^3 dependence of $\langle u^2 \rangle$, described in Sect. 2.2. The *dashed* error bars are associated with PIMC points (□) also given.

perature makes it easier to see whether a ratio of $\sqrt{4/3}$ is plausible. In Fig. 9 the dashed line comes not from an empirical relation, but rather from PIMC computations described in Sect. 2.2.

2.2 Comparison to PIMC Computations

As stated in Sect. 1 and shown in Table 1 of Sect. 3, below, excellent agreement is achieved between precisely–measured kinetic energies, $\langle E_k \rangle$, and PIMC computations using sample sizes as small as 256 atoms [1,4].

Therefore, it was initially surprising, when precise values of $\langle u^2 \rangle$ were measured [7,8], that these showed systematic differences above PIMC values using similar small sample sizes. These differences, far larger than the sum of the respective errors of the measured and of the PIMC values, were resolved by Draeger and Ceperley [9]. They devised efficient methods of carrying out computations of $\langle u^2 \rangle$ for large sample sizes, N, and indeed extrapolated each computation to the thermodynamic limit $1/N \rightarrow 0$, to obtain a final value they denote $\langle u^2 \rangle_\infty$. Figure 10 shows a typical result. One sees that the extrapolation in this case has produced an increase in the computed $\langle u^2 \rangle$ value of 16%. Agreement with the measured value at the limit [7] is excellent.

The form of the extrapolation is not obvious. Indeed, understanding the dependence upon $1/N^\alpha$, that is, that the value of $1/3 < \alpha < 2/3$, led to the

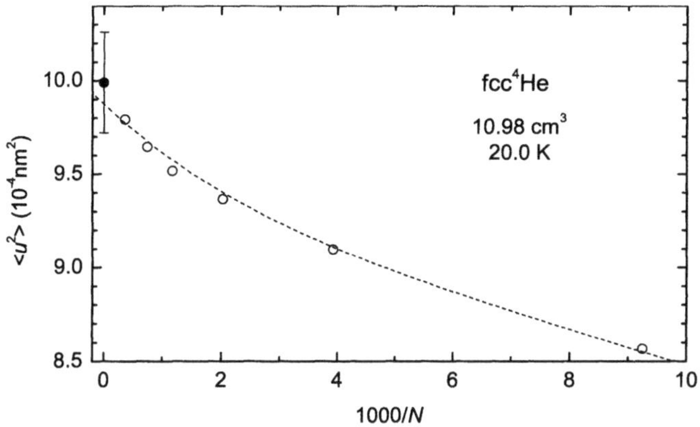

Fig. 10. Extrapolation to the thermodynamic limit $N \to \infty$ is required in order to obtain agreement of the PIMC simulation (∘) with precise experimental measurements of $\langle u^2 \rangle$ in helium crystals [9]. This example applies to fcc ^{4}He at molar volume 10.98 cm^3 at 20.0 K. The experimental data point (•) is from Refs. [7,40].

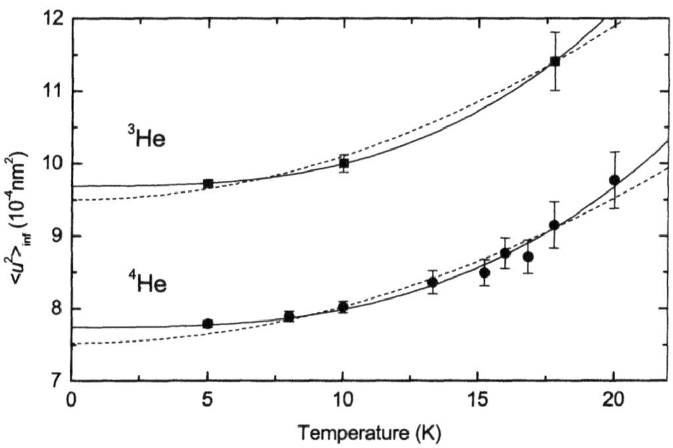

Fig. 11. PIMC computations of $\langle u^2 \rangle_\infty$ [9] suggest that T^3–dependencies are indicated (*solid lines*), contrary to a T^2–dependence expected in a harmonic approximation (*dashed lines*). The PIMC results shown are for fcc ^{3}He at $V = 11.54$ cm^3 and fcc ^{4}He at 10.98 cm^3, respectively.

conclusion that in the range of V's and T's studied, the He samples are in a regime of crossover between classical and quantum behavior.

Once a satisfactory extrapolation was found to obtain values for $\langle u^2 \rangle_\infty$, other consequences were explored, beyond the reach of experiments. An interesting example is the T–dependence of $\langle u^2 \rangle_\infty$, as shown in Fig. 11 for both isotopes. In the harmonic approximation, one expects a T^2 dependence. However, in both isotopes a T^3 dependence appears more satisfactory in the available range up to melting. The harmonic approximation is not satisfactory, even in some 'effective'

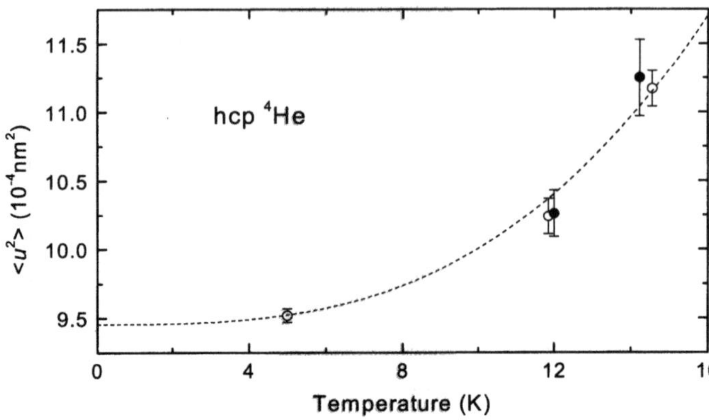

Fig. 12. Comparison of PIMC computations of $\langle u^2 \rangle_\infty$ [9] (∘) and measurements [7] (•) for 12.12 cm^3/mol hcp ^{4}He. Agreement is excellent, but the measurements neither confirm nor deny the T^3-dependence shown.

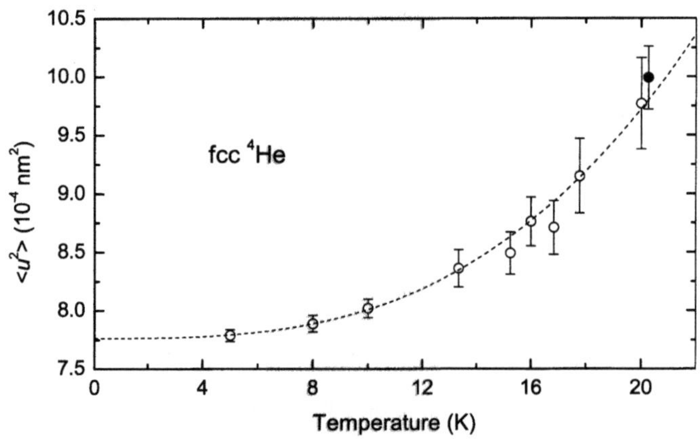

Fig. 13. Comparison of PIMC computations of $\langle u^2 \rangle_\infty$ [9] (∘) and measurement [7] (•) for 10.98 cm^3/mole fcc ^{4}He. There is only a narrow range of fcc available for measurements at this molar volume (Fig. 4). Note, the T-variation of $\langle u^2 \rangle_\infty$ is comparable to the difference produced at fixed T by PIMC extrapolation, Fig. 10.

form, to describe the dependence upon T. This matter is discussed a bit more below in the language of the Debye characteristic temperature, Θ_M.

The question of agreement between PIMC $\langle u^2 \rangle_\infty$ values and experiment as shown in Fig. 10 is paramount. Additional examples are illustrated in Figs. 12 and 13. These show PIMC $\langle u^2 \rangle_\infty$ values and experimental data for hcp ^{4}He at molar volume 12.12 cm^3 and fcc ^{4}He at 10.98 cm^3, respectively. For hcp the data agree excellently, but there are not enough measured values to confirm or deny a T^3 behavior. For fcc there is also excellent agreement at the single experimental

T. It is unfortunate that fcc phase in nature is so limited in extent at this molar volume.

As already shown in Fig. 9 displaying Θ_M, the question of T–dependence can be approached empirically, independent of specific simulations, given an assumed Grüneisen V–scaling. As a matter of interest, the results from PIMC simulation can also be represented in this manner [7]. The dashed curve is that from a fit of the form $a + bT^3$ to the fcc ^{4}He PIMC computation for a molar volume 10.98 cm^3 already shown directly in $\langle u^2 \rangle$ in Fig. 13. This shows a reflex behavior with a maximum, but does appear to approach $T \to 0$ with zero slope, as one would expect thermodynamically. It should be noted that such a maximum is not present in Θ_M at constant V in a harmonic approximation. The harmonic case has a low–T expansion which is monotonic increasing from $T = 0$, and a high–T expansion which from its high–T limit is monotonic and decreasing with decreasing T [34]. At present there is only one measurement below the possible maximum in Θ_M, which occurs near 10 K at this molar volume. That limited measurement [47] consisted of intensity data for fixed T at only three values of Q and, at least as far as the present scheme of volume corrections is concerned, lies far off the ^{4}He dashed curve in Fig. 9.

Next, consider the isotopic differences shown on Fig. 9. The close agreement between recent measurements [7] on ^{4}He and PIMC simulation has just been noted. In addition, to represent a first approximation to isotopic effect, one might put another curve, produced by a simple mass–scaling factor of $\sqrt{4/3}$ applied to the ^{4}He PIMC simulation. This is a frequency scaling $\hbar\omega = k_\mathrm{B}\Theta_M$ at the same V. The present ^{3}He measurements fit this plausibly. Exceedingly precise X–ray work at lower T's would be required to be more discriminating.

Finally, it should be noted that the PIMC computations also produced a small anisotropic Q^4 term in the Debye–Waller factor schematically represented in (5). Such a term indicates a computed non–Gaussian distribution of the atoms around their lattice sites. In measurements up to now, however, neither anisotropy nor non–Gaussian behavior has been detected in He crystals.

3 Inelastic Scattering

The extension of scattering probes to measure not only $\boldsymbol{Q}$–dependent information about a condensed matter system but also the inelastic response has brought about marvelous extension of knowledge about the dynamics of such systems. One arranges to measure the double differential cross–section [49],

$$\frac{\mathrm{d}^2\sigma}{\mathrm{d}\Omega\mathrm{d}\omega} = \frac{1}{4\pi}\left(\frac{k_f}{k_i}\right)\sigma_{\mathrm{coh}}S_{\mathrm{coh}}(\boldsymbol{Q},\omega) + \frac{1}{4\pi}\left(\frac{k_f}{k_i}\right)\sigma_{\mathrm{inc}}S_{\mathrm{inc}}(\boldsymbol{Q},\omega)\,, \tag{11}$$

where $S(\boldsymbol{Q},\omega)$ is the dynamic structure factor, which in general has both coherent and incoherent parts for neutrons. Extensive discussion of this method for neutron probes appears elsewhere in this book [50].

Neutron probes have some advantages which over a half century have been highly developed. Among them are that extremely high resolution is available in energy (ω), that thermal neutron wavelengths are similar to typical interatomic distances in condensed matter, and that the coherent and incoherent cross–sections, σ_{coh} and σ_{inc}, respectively, are different for each isotope, so that it may be possible selectively to highlight particular aspects of the dynamics under study. On the other hand, in a few cases, such as ^{3}He, nuclear absorption is so large as to make use of a neutron probe difficult and inaccurate, owing to limited counting statistics. For this reason, practically all of the myriad dynamical studies on condensed helium have been on the boson ^{4}He [23].

As mentioned in Sect. 1, neutrons of eV energies can be used in recoil experiments to obtain the momentum distribution of the target nuclei. Neutron beams at such energies are obtained from a pulsed spallation source, through partial moderation. In the limit of the impulse approximation, the dynamic structure factor becomes

$$S_{\mathrm{IA}}(\boldsymbol{Q},\omega) = \int_{-\infty}^{\infty} n(\boldsymbol{p})\delta\Big(\omega - \omega_r - \frac{1}{m}\boldsymbol{Q}\cdot\boldsymbol{p}\Big)\mathrm{d}\boldsymbol{p}\,, \tag{12}$$

where the momentum distribution $n(\boldsymbol{p})$ is normalized to unity, $\hbar\omega_r$ is the recoil energy $E_r = \hbar^2Q^2/2m$, and the δ function denotes conservation of momentum and energy in each collision. According to this, impulse approximation scattering will yield a profile of $n(\boldsymbol{p})$ along the direction of $\boldsymbol{Q}$ for the collection of target nuclei, considered as independent scatterers.

In many condensed noble gases and in crystal and fluid H_2, a Gaussian profile has been found sufficient to represent measurements made up to now, and the kinetic energy, $\langle E_k\rangle$ has been determined at various T's. PIMC simulations have found a weak exponential tail in normal fluid ^{4}He [31], but non–controversial evidence for this from measurements has been lacking so far. Of course, superfluid ^{4}He has a different $n(\boldsymbol{p})$, but that subject is outside the scope of this chapter.

In condensed He, $\langle E_k\rangle$ is dominated by ground–state energy at low temperatures, where different phases at the same molar volume are accessible through different imposed conditions of P and T. It is therefore of interest to see whether there are differences in $\langle E_k\rangle$ for these phases. Some might be expected because of different degrees of local spatial order in the phases. Table 1 shows the first results [3]; refinement and elaboration of the story appears in [4]. The V chosen is shown in Fig. 2 by symbol $\diamond$. The overall conclusion is that there is little difference between $\langle E_k\rangle$ for bcc, hcp, and normal fluid phases at constant molar volume, and that effects of differing T are minimal at these low T's. When the results in Table 1 are examined closely, however, they show that $\langle E_k\rangle$ for fluid ^{4}He is somewhat greater than that of solids at the same density. Apparently each atom has somewhat more neighborhood space in the crystalline structures.

This kind of work is being extended to ^{3}He, in which neutron absorption is severe [51]. Related path-integral computations have been published [28].

In some sense, the position– and momentum–coordinate quantities $\langle u^2\rangle$ and $n(p)$ are complementary. In other senses they are not. The quantity $\langle u^2\rangle$ is accessible only in crystalline phases; it is dependent upon averaging over a well–defined lattice, hence its sensitivity to sample size as noted in Sect. 2.2. On the other

Table 1. First report of kinetic energies, $\langle E_k \rangle$, of different ^{4}He phases all at the same molar volume [3]. $\langle E_k \rangle$ is approximately constant in different phases, but $\langle E_k \rangle$ in liquid is slightly higher than $\langle E_k \rangle$ in solid. A comprehensive summary over boundaries in the phase diagram appears in Ref. [4].

Phase	T (K)	V (cm^3/mol)	$\langle E_k \rangle$ (K)	Method
liq	2.705	20.91	24.2	Expt
liq	2.353	20.90	24.8	PIMC
bcc	1.725	20.91	23.7	Expt
bcc	1.667	20.90	24.1	PIMC
hcp	1.070	20.91	23.6	Expt
fcc	1.600	20.90	24.4	PIMC

hand, the quantity $\langle E_k \rangle$, obtained from $n(\boldsymbol{p})$ by a spherical averaging, is dependent primarily upon the local environment, or rather, the number density of the local environment and relatively short-range correlations. The PIMC computations which demonstrated significant differences between $\langle u^2 \rangle$ and $\langle u^2 \rangle_\infty$ showed little if any differences between the corresponding kinetic energies [9]. Finally, a search for anisotropy of $\langle u^2 \rangle$ in hcp ^{3}He did not detect any [7]. Precise scattering studies are not yet available in crystal phases at constant number density, in order to check for possible small differences in average $\langle u^2 \rangle$, analogous to those displayed for $\langle E_k \rangle$ in Table 1. PIMC computations show that little difference between fcc and hcp phases is to be expected.

For the case of a photon probe (11) becomes

$$\frac{\mathrm{d}^2\sigma}{\mathrm{d}\Omega \mathrm{d}\omega} = r_0^2 (\boldsymbol{\varepsilon_i} \cdot \boldsymbol{\varepsilon_f})^2 |f(\boldsymbol{Q})|^2 S_{\mathrm{coh}}(\boldsymbol{Q}, \omega) \;, \tag{13}$$

where r_0 is the classical radius of the electron, the $\boldsymbol{\varepsilon_i}$ and $\boldsymbol{\varepsilon_f}$ are incident and scattered photon polarizations, respectively, and $f(Q)$ is the atomic scattering factor [6]. For He, $f(Q)$ is small and moreover has a severe Q–dependence. This, together with the difficulty of achieving useful energy resolution with X–rays, prevented any direct dynamical studies using photons on condensed He until the second generation of synchrotron sources could be applied [52].

The measured intensity of photon scattering is given by

$$I(\boldsymbol{Q}, \omega) \propto \int_{-\infty}^{\infty} (a\delta(\omega') + S(\boldsymbol{Q}, \omega))\, R(\omega - \omega')\, \mathrm{d}\omega' \;. \tag{14}$$

In this expression, the delta function denotes elastic scattering processes, which in solid He work mostly arise from the polycrystalline Be sample cell required to confine the sample at a given P. This elastic peak is clearly identified and characterized in empty–cell scans, made at the same individual orientations of the sample cell used in the corresponding inelastic scans with He in the

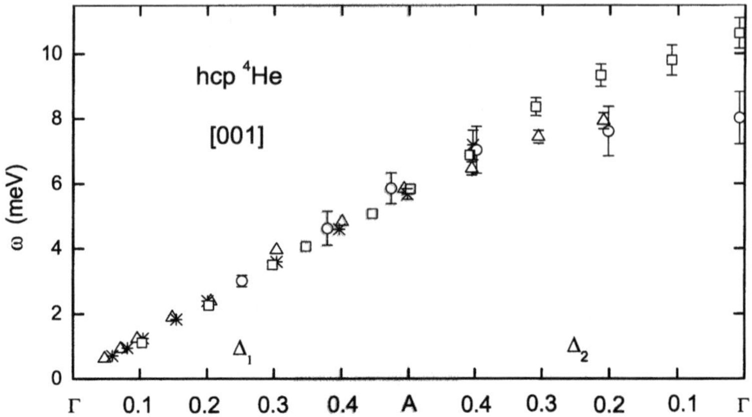

Fig. 14. All known longitudinal branch dispersion data from inelastic neutron scattering, displayed in the reduced–zone scheme, in hcp ^{4}He in the [001] direction, Δ_1 and Δ_2. Results were scaled, using (10), to a molar volume of 13.25 cm^3, from experiments on samples at these molar volumes: 9.4 and 11.6 cm^3 [54] (*triangles* and *stars*, respectively); 16.0 cm^3 [58] (*boxes*); 21.1 cm^3 [56] (*circles*).

cell. The elastic scattering can be used to obtain the location of zero energy transfer in each individual scan of the spectrometer. One great advantage of triple–axis X–ray technique is that the spectrometer resolution, $R(\omega)$, is independent of the momentum transfer, unlike the case for neutron spectrometers which require more elaborate corrections. Further, for X–rays it is also often possible to reduce instrument backgrounds to negligible levels compared to the weak inelastic scattering, again unlike the neutron case. Both these latter advantages make it possible to use X–rays to measure relative intensities over a very large range of energy transfer, $\hbar\omega$, with an assurance which encourages determination of phonon energy eigenvalues and eigenvectors of modes in complex structures [6,53].

3.1 Neutron Studies of Phonons in Helium

Information on phonons in various phases of ^{4}He was produced in a series of investigations at Brookhaven, on hcp, bcc, and fcc [54–57], and at Iowa State, on hcp [58]. This body of work stimulated intensive development of models, in particular those of self–consistent–phonon type [23,29,30]. One of the basic issues was that of density scaling. Particularly for acoustic modes, an expression like (10) was found to be satisfactory, even though (10) depends on an average over the phonons as taken in heat capacity measurements. This is illustrated in Fig. 14, which gives data from the longitudinal $\Delta_{1,2}$ branch in hcp [001] direction. The scaling is made to a molar volume of 13.25 cm^3, which is shown by symbol $\triangle$ in Fig. 2. Data are shown from samples at a very wide range of molar volumes, from 9.4–21.1 cm^3 [54,56,58]. In all these data, the energy was determined at a given $\boldsymbol{Q}$ by examining the position of the peak S in the energy response.

For a solid exhibiting modest amounts of anharmonicity, it may be appropriate to expand the coherent dynamic structure factor S. If $\langle(\boldsymbol{Q}\cdot\boldsymbol{u})^2\rangle$ is small at low $\boldsymbol{Q}$, then symbolically:

$$S(\boldsymbol{Q},\omega) = S_1(\boldsymbol{Q},\omega) + S_2(\boldsymbol{Q},\omega) + S_{12}(\boldsymbol{Q},\omega) + S_{13}(\boldsymbol{Q},\omega) + \dots \; . \tag{15}$$

In this expansion, the terms S_1, S_2, etc. represent successive orders of a harmonic expansion, while the terms like S_{12} represent interactions between one– and two–phonons, etc. Effects owing to the interference term S_{12} have been identified in neutron scattering spectra from ^{4}He, because they leave a signature when the scattering is measured across a Brillouin zone boundary [23]. This signature arises because $S_{12}(\boldsymbol{Q},\omega)$ is proportional to Q^3, which changes sign for $\boldsymbol{Q}$ in adjacent zones, i.e. it changes sign at $\boldsymbol{q} = \boldsymbol{Q} - \boldsymbol{\tau}$, where $\boldsymbol{\tau}$ denotes the zone boundary. If $S_1(\boldsymbol{Q},\omega)$ is sharply peaked, then the effect is to change the intensity of the observed $S(\boldsymbol{Q},\omega)$. If on the other hand $S_1(\boldsymbol{Q},\omega)$ is broad, then the effect will also shift the location of the observed response. Such effects may be responsible for the scatter seen in Fig. 14 for the Δ_2 branch; there the shapes of the scattering response evidently made determination of the peak position variable.

Later in discussing X–ray scattering results in the context of sum rules, a different segregation of the expansion (15) into two parts will be used:

$$S(\boldsymbol{Q},\omega) = S_p(\boldsymbol{Q},\omega) + S_2(\boldsymbol{Q},\omega) + \dots \; . \tag{16}$$

Here, S_p is the sharp component, the sum of one– and of one–multi–phonon–interference terms [59].

The early SCP calculations on He were successful in important ways, which included their obtaining phonon frequencies ω^2 which were positive, as opposed to the negative squared values for crystal He given by traditional harmonic methods [60].

However, these early calculations have both known limitations in principle [37] and quantitative deficiencies in practice. Consider, for example, the situation with respect to predicted dependencies of the phonon frequencies upon V. Calculations were made in the range 16–21.5 cm^3/mol and in the range 10–16 cm^3/mol. The paper at larger V's [61] noted that this was the least satisfactory aspect of that work, although the phonon frequencies at any particular V in the range did not drastically disagree with experiment. Figure 15 shows the calculated phonon frequencies of the longitudinal $\Delta_{1,2}$ phonons in the [001]–direction at hcp zone point A, (0, 0, 0.5), compared to neutron data [54,56,58]. The paper discussed possible sources of the discrepancy, including the treatment of short-range correlations and the absence of any phonon damping and associated frequency shifts. The paper at smaller V's [62] aimed to minimize the relative importance of short-range correlation approximations by treating higher densities, where the static lattice energy is relatively larger compared to the vibrational energy. It also investigated possible isotopic mass effects. These gave a

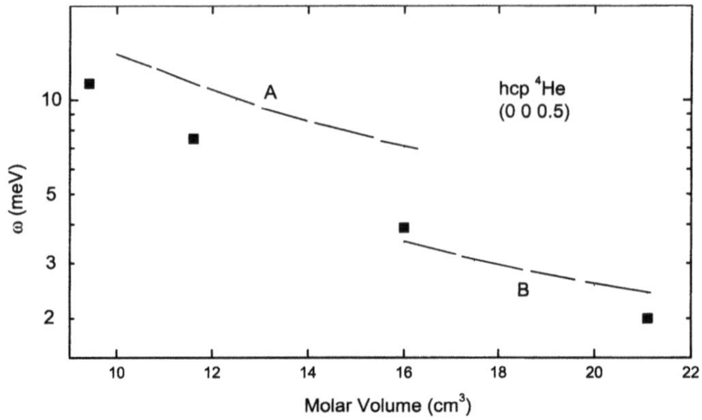

Fig. 15. Existing SCP theories for hcp He (*curve* A [62], *curve* B [61]) exhibit wrong density dependencies. Shown is the longitudinal phonon energy at (0,0,0.5). Moreover, each yields values different from those determined from experiments at the relevant molar volume [11]. Neutron data shown (*black boxes*) are from Refs. [54,56,58].

phonon ratio 1.23, larger than $\sqrt{4/3}$, and outside the error limits of X–ray work on isotopes discussed below. Again, the calculated V-dependence is only about one-half of that observed. The subject of density dependence has hardly been visited in any comprehensive way since that time.

The early literature noted that the calculated phonon frequencies as measured typically disagreed with results based on various implementations of SCP theory by about 20%, but discrepancies of this magnitude were not regarded as particularly disturbing. Unfortunately, even with the publication of further theoretical work, this modest degree of satisfaction with the situation became usual. Rather, the focus of the discussion shifted to other issues such as excitation line shapes, attributed to phonon interferences [2,23].

Even in fcc ^{4}He at 9.03 cm^3/mole, that is, even more compressed than point (□) in Fig. (2), the comparison between theory [63] and experiment [57] is not impressive. This small V might represent a case where $\langle u^2 \rangle$ values are small enough to encourage use of the expansion (15), so that calculation of specific phonon interferences might be verified by experiment. Three such longitudinal phonon groups along the [111] direction are shown in Fig. 16. It can be seen that there is severe background from the massive pressure cell required for the large sample for neutron work. This precluded confirmation of the predicted line shapes, and therefore precluded confirmation that the specific phonon interferences considered are dominant.

3.2 High–Resolution X–Ray Inelastic Studies

The earliest reported meV–resolution X–ray work on condensed He included response from both solid and fluid [52]. Some subsequent X–ray work on fluid He has been reported [64]; this concerned quantum effects in the dynamics of supercritical fluid ^{4}He. The claim is made that even in the regime of collective

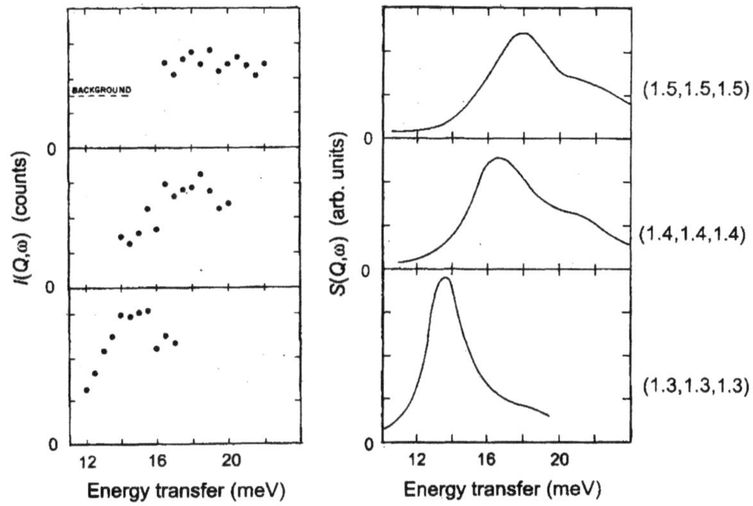

Fig. 16. A published comparison for L[111] phonons in fcc ^{4}He at 38 K and 9.03 cm^3/mol between experiment [57] and theory [63]. Estimated neutron background is shown on panel at upper left. Confirmation of theory is lacking.

response at low Q, it is possible to extract some single–particle information such as $\langle E_k \rangle$. Here, discussion will focus upon how X–ray work has established continuity with the body of previous neutron study of solid ^{4}He, and has extended it with new results on ^{3}He [65].

Coherent inelastic X–ray scattering work on crystal He [10,11] used as a fitting function a convolution of a Lorenzian $R(\omega)$, representing the instrument resolution, and an expression for a damped harmonic oscillator (DHO):

$$S_{\text{fit}}(\boldsymbol{Q},\omega) = \frac{Z_q(\boldsymbol{Q})}{1-\exp\left(-\hbar\omega/k_{\text{B}}T\right)} \frac{\omega\omega_q\Gamma_q}{(\omega^2-(\omega_q^2+\Gamma_q^2))^2+4\omega^2\Gamma_q^2} \,. \tag{17}$$

In this expression, $Z_q(\boldsymbol{Q})$ is the phonon structure factor, and the denominator below it takes account of detailed balance between the energy–loss and energy–gain processes in the X–ray scattering. Γ_p is a damping width (in frequency units) associated with the excitation of frequency ω_q; it vanishes in the harmonic approximation. An example of the inelastic X–ray response is shown in Fig. 17, for hcp ^{3}He at a molar volume of 13.25 cm^3. Incident X–ray energy was 17.79 keV obtained by elaborate monochromatization of undulator radiation from the ESRF Grenoble and the response was measured with 5.9 meV resolution. The elastic line comes from the Be pressure cell; the inelastic scattering from Be lies well outside the range of energies scanned for He phonons.

The deduced dispersion of the longitudinal Δ_1 and Δ_2 branches in hcp ^{3}He is shown in Fig. 18. Data were taken in an extended range, not only from Γ to A to Γ (the first Δ_1 and Δ_2 ranges), but also in the next (the Δ_2 and Δ_1 ranges again, in reverse order with increasing Q). For comparison, the dashed line represents a monatomic linear chain model. One sees apparently systematic deviations from

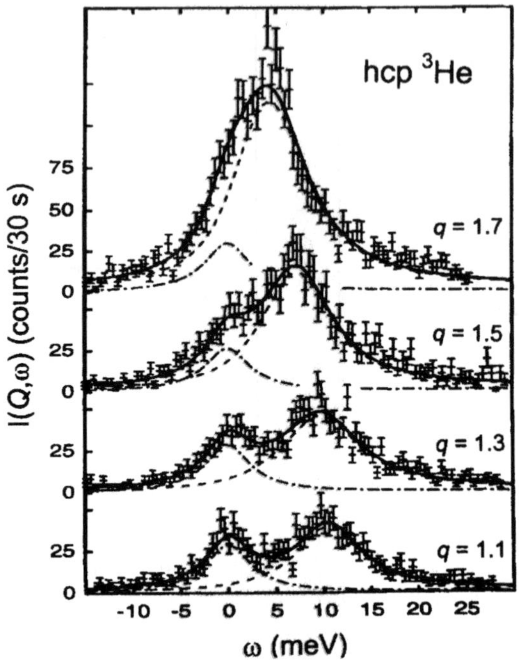

Fig. 17. Inelastic X–ray response of hcp ^{3}He at 12 K as measured by Seyfert *et al.* [11]. The wave vector transfer was measured along the [001] direction. The zero of energy transfer is defined by the elastic scattering of the Be cell (Fig. 5) and resolution broadening is included.

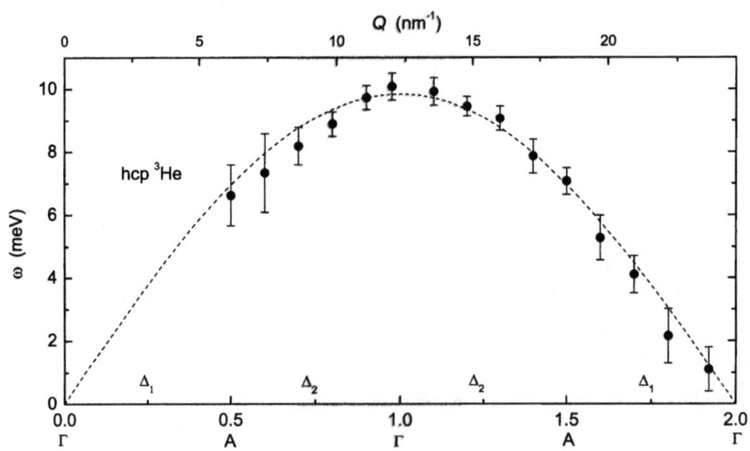

Fig. 18. Dispersion of the phonon response in hcp ^{3}He as measured in extended zones by X-ray inelastic scattering [11]. The indicated uncertainty in the region near $Q = 15\ \mathrm{nm}^{-1}$ is about 0.4 meV, which corresponds to a precision of about 2×10^{-8}. The *dashed* line represents a monatomic linear chain model.

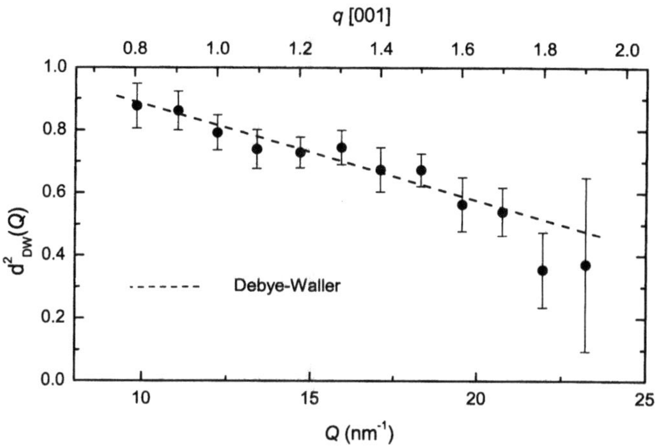

Fig. 19. The intensities of c–axis phonons in hcp ^{3}He as measured by Seyfert *et al.* [11]. The *dashed* line shows the result expected from direct Debye–Waller work [7].

this simple model, which are likely indications of the phonon interference effects mentioned in Sect. 3.1. When these X–ray data are compared to Fig. 14, they agree well with neutron data, to the extent, already noted, that the neutron data are agreeable with each other.

It is notable that even in this early stage of development of X–ray inelastic scattering from He at meV resolution, useful intensity measurements can be obtained. Figure 19 shows an example, again for ^{3}He [11]. It can be seen that there is good agreement with the Debye–Waller factors already discussed [7].

In a further paper Seyfert et al. [66] report on longitudinal *c*–axis excitations in hcp ^{4}He in the same arrangement, and at about the same crystal density. These were similarly analyzed with a damped harmonic oscillator (DHO) model over an extended range. When multi–phonon processes are present, a DHO model is an imperfect description, but the consistency of a DHO model for work at currently available spectrometer resolution was empirically tested [11]. Now focus on the extraordinarily large values of Γ_q found in the scattering experiments from both ^{3}He and ^{4}He [10]. In Fig. 20 is shown the damping parameter, $2\Gamma_p$, in a DHO model fitting these *c*–axis longitudinal excitations in ^{3}He at 13.25 cm^3/mol, taken using 17.79 keV X–rays (solid circles). Also shown (open circles) are data for ^{4}He at 13.2 cm^3/mol, taken using 13.84 keV X–rays. In neutron work on ^{4}He near the same density, excitation broadening was seen in the Δ_2 branch, and a figure shows an apparent abrupt change, near zone point A, from no broadening in the Δ_1 branch [58]. Unfortunately, the paper does not specify the particular extended zones scanned to obtain these results, so one can only speculate about the possible Q dependence.

In both cases, ^{3}He and ^{4}He, the damping is extraordinarily greater in the q range from 1 to 2 than from 0 to 1. This is contrary to expectation in a one–phonon picture that these should be equivalent. Various possibilities were examined for these results. For example, if the structure is not strictly hcp, there will

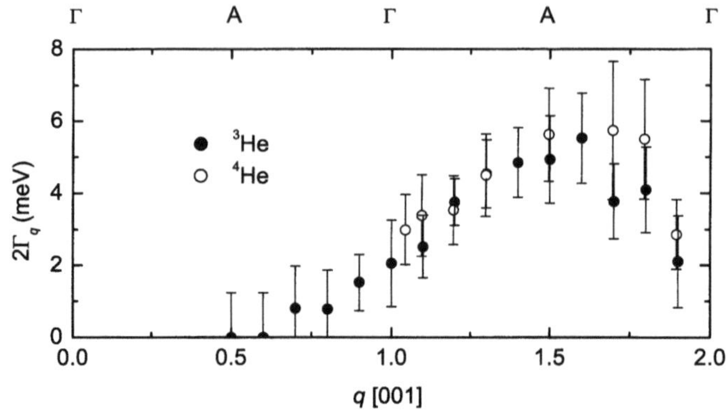

Fig. 20. Phonon excitation widths grow remarkably broad as X–ray inelastic measurements are extended into distant zones. At the second A point, the width is comparable to the energy. Compare Fig. 18. Data for both hcp ^{3}He and ^{4}He are shown [10,11].

be a Bragg peak, albeit weak, at (001). An experimental search revealed no such reflection. Indeed, the very extensive Debye–Waller work of Arms [7] on many additional hcp He crystals grown in the same kind of Be sample cell (Fig. 5) found only those Bragg peaks with (hkl) values of hcp. This is contrary to a published expectation from SCP theory [67]. If the crystals, grown in confined conditions, had many stacking faults, there would have been some other peaks present [68]. While one could conclude that multi–phonon excitations are responsible and leave it at that, it would seem fruitful to pursue further investigation, both in experiment and in theory.

The presence of multi–phonon contributions can be investigated, at least qualitatively, by calculating the first moment of the sharp component of (16), namely $S_p(\boldsymbol{Q},\omega)$. According to the ACB sum rule [59], this is

$$\int_{-\infty}^{\infty} \omega S_p(\boldsymbol{Q},\omega)\mathrm{d}\omega = |G(\boldsymbol{Q},\boldsymbol{q})|^2 \mathrm{d}^2{}_{DW}(\boldsymbol{Q}) \, . \tag{18}$$

In hcp ^{3}He along the c–axis direction, the square of the inelastic structure factor $|G|^2$ is just $Q^2/2m$. Figure 21 shows M1, the left–hand side of (18), divided by $Q^2/2m$, and also the Debye–Waller factor from direct measurements, plotted versus Q. It is interesting to see that these differ from one another exactly in the region $1 < q < 2$ where the excitation width Γ_p is broadened (Fig. 20). This region is therefore one with multi–phonon effects, beyond those involving single phonons.

These different lines of evidence, for massive multi–phonon contributions in He, once again emphasize the deficiencies of existing SCP models which limit their considerations to selective numbers of processes and to inadequate treatment of short-range correlations. Even more comprehensive theoretical work would be helpful.

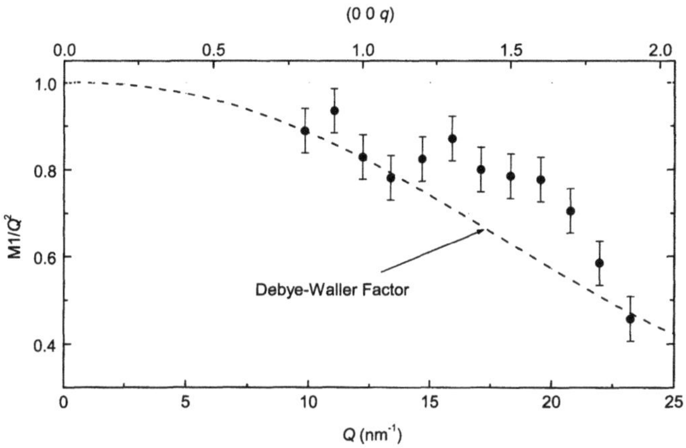

Fig. 21. The ACB sum rule gives a test for multi–phonon effects, where the difference from the Debye-Waller factor is easily seen.

4 Equilibrium Defects

Helium crystals have long been recognized as a medium for the possible occurrence of point and line defects of unusual character. As verified by the Debye–Waller results already described in Sect. 2, the crystals have extraordinary amplitudes of atomic oscillation. As shown by NMR studies of ^{3}He and of isotopic mixtures, there are atomic tunnelling phenomena. Finally, there are possibilities arising from both these that the statistics of the atoms can influence the defect structures, because of the relative ease of atomic exchange, even at the lowest temperatures [24].

There is no space here for a comprehensive review of experimental results in He solids. One durable paradox is the apparent contradiction between direct measurement of thermal vacancy content by X–ray diffraction and indirect inferences of that content from measurements of other types. Those inferences have led to greatly differing estimates, different both from the X–ray work and from each other. Ordinarily, one would expect that direct structural information would be persuasive. However, such measurements are quite difficult, especially at temperatures below 1 K. Only one group has reported extensive X–ray work, and for historical reasons in other difficult areas of research, the low–temperature community has been reluctant to accept even direct results without confirmation by an independent group. For a brief summary of some aspects of this situation in solid He, see Simmons [16] and the many references therein. Among subsequent papers on specific topics are research reports on anisotropy in ion transport [69], on self–diffusion and related atomic mobility [70], and on high sensitivity pressure measurements searching for ground state vacancies [71]. Controversial interest persists in several other areas, for example on ground state issues [72]. Below, some recent additional X–ray work is described and discussed.

4.1 Some Definitions and Relations

By consideration of the total system consisting of a hypothetical perfect crystal plus the defects, one can separate partial defect quantities from the thermodynamic functions for the total system. Among the considerations is an appropriate definition of the configurational entropy of the defects which in a classical system are localized and are randomly distributed upon lattice sites. It can then be arranged that relations among the partial defect quantities look the same as usual thermodynamic relations [73].

Thermal defects in solid He are typically studied in a vessel which is rigid compared to easily compressible He. The appropriate free energy for the system is then the Helmholtz energy, $F(V,T)$. In such a case the suitable defect free energy of formation for these defects is also Helmholtz. For example, for monovacancies

$$f(V,T) = u(V,T) - Ts(V,T) \tag{19}$$

where $u(V,T)$ is the energy of formation and $s(V,T)$ is the non–configurational entropy of formation. The equilibrium vacancy concentration at this molar volume, V, is then

$$c = \exp\left(-f(V,T)/k_{\mathrm{B}}T\right) . \tag{20}$$

The corresponding partial pressure of vacancy formation is

$$p(V,T) = -\left(\frac{\partial f}{\partial V}\right)_T . \tag{21}$$

From the partial Gibbs energy of formation $g(P,T)$, the partial free volume of vacancy formation is given by

$$v(P,T) = -\left(\frac{\partial g}{\partial P}\right)_T . \tag{22}$$

At a given P, V, and T, equilibrium vacancy content c is independent of whether the state was reached under isobaric or isochoric conditions or some other path, and hence there $f = g$. At this state one therefore has

$$p = -\left(\frac{\partial g}{\partial P}\right)_T\left(\frac{\partial P}{\partial V}\right)_T = -v\left(\frac{\partial P}{\partial V}\right)_T = \frac{v}{V}B_T , \tag{23}$$

where B_T is the isothermal bulk modulus.

A useful empirical relation involves the isothermal bulk modulus. It is

$$g = CB_T\,v_a , \tag{24}$$

where C is a constant and v_a is the atomic volume [74]. It will be applied in Sect. 4.3 below.

It should be remembered that the thermodynamic relations above hold when the defects are localized, as they are in usual solids. However, in some He solids the defects may be non–localized, and the definition of defect entropy may require

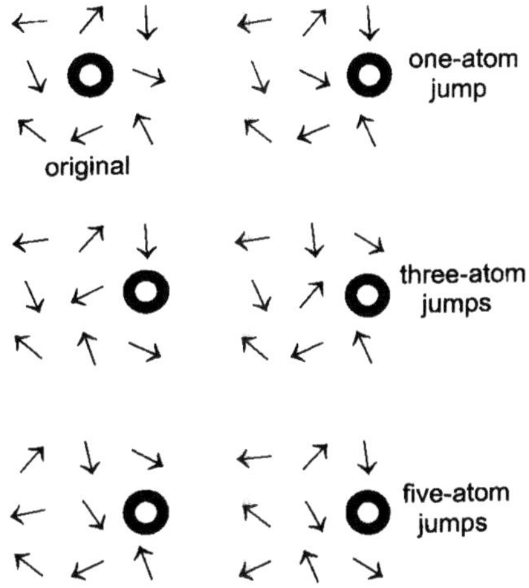

Fig. 22. Schematic representation of several different possible final states after motion of a vacancy in ^{3}He, in each case from the original state shown at upper left. At temperatures above the magnetic transition, the nuclear spins are randomly oriented, so that motion of a vacancy can occur by jumping of adjacent atoms in multiple ways, each of which produces a different final state for the system (see text). The description of the vacancy state therefore must include these independent possibilities.

re-examination. Such cases have been hypothesized to be present, for example, under conditions at extreme low temperatures and low density. To illustrate a possible condition, consider the case of ^{3}He, which has been extensively studied [15,75]. In Fig. 22, the site at which the vacancy is centered is represented, schematically in two dimensions, by the heavy circle. In the first row, specification of the initial state includes for example, the orientation of the nuclear spin of each of the atoms. Because the ordering temperature of these spins is only a few mK, they are shown to be randomly disordered. For simplicity in the illustration, shown only is the site of the vacancy and its near neighbors in a plane. At upper left, in the top row, the initial state is shown. The other five diagrams show some of the different possible final states after a vacancy move to its near-neighbor site to the right.

At the right of the top row a possible final state is shown, corresponding to the hopping of one atom from the right to the left (hopping of the vacancy from the left to the right). A little consideration shows, however, that such a hopping of the vacancy such a distance to the right can also be effected by multiple hops of surrounding atoms. The second row shows two such rearrangements involving successive motions of three atoms. Note that in the two cases, the final state is different because the resulting array of spin orientations is different. The bottom row shows vacancy hops requiring five atoms. Of course, in three dimensions, the number of possible 3–atom jumps is much enlarged. Such cooperative motions are the origin of the interesting magnetic properties of non–defective ^{3}He crystals [76], for which the hops are tunnelling interchanges of pairs–, of triplets–, etc., of atoms.

In quantum solid ^{3}He such vacancies indeed move by means of tunnelling processes; the NMR evidence shows at low T, a T–independent diffusion [12]. The "tunnelling vacancy" state can therefore be represented, not simply by specifying which atomic site is unoccupied, but by forming the vacancy state as a linear combination of states:

$$\Psi_v = \sum_\alpha a_{v\alpha} \psi_{v\alpha} \, . \tag{25}$$

There is in such a representation an entropy problem, which has been considered, for example, by Hetherington [15]. Understanding it may help resolve the "durable paradox" noted at the beginning of this section.

4.2 Direct Measurement of Net Vacancy Content

The most direct measure net vacancy content, in a solid composed of identical units (atoms or molecules), is

$$c = \frac{\text{net added number of substitutional sites}}{\text{total number of units}} \, .$$

This is implemented through the simultaneous determination of macroscopic and of diffraction molar volume differences, preferably through reversible measurements as the temperature is cycled.

$$c(T) = \frac{\Delta V(T)}{V(T)} - \frac{\Delta V_X(T)}{V_X(T)} \, . \tag{26}$$

The method is simply a comparison between the number of units and the number of unit sites. It is independent of lattice relaxation in the neighborhood of the defects, because contributions of such relaxation to the respective macroscopic and diffraction measurements cancel in the subtraction. Note further, this method is independent of localization of the defect, so it is applicable even in the extreme situation illustrated in Fig. 22 and represented by (25). There is no uncertainty in this independence, corresponding to the possible uncertainty noted in definition of the partial defect entropy.

Because some net defect concentrations of interest are likely to be very small ($c < 10^{-5}$), there is a limitation in practice with the sensitivity of the diffraction method. Fortunately, unambiguous results have been obtained for several He systems [16]. In metals, a favorable case is fcc Al, which is metallurgically stable owing to its oxide skin. Many independent diffraction and associated bulk studies agree on the vacancy content in pure Al. Other favorable cases include certain metallic alloys exhibiting ordering behavior and large thermal vacancy content [77]. For these there is an extraordinary analogy in He isotope mixture solids (Sect. 4.4). For the metal prototype pure Cu, however, there are only two diffraction studies which agree [78,79]. For Cu and for other pure metals for which the vacancy content is small and the experimental conditions adverse,

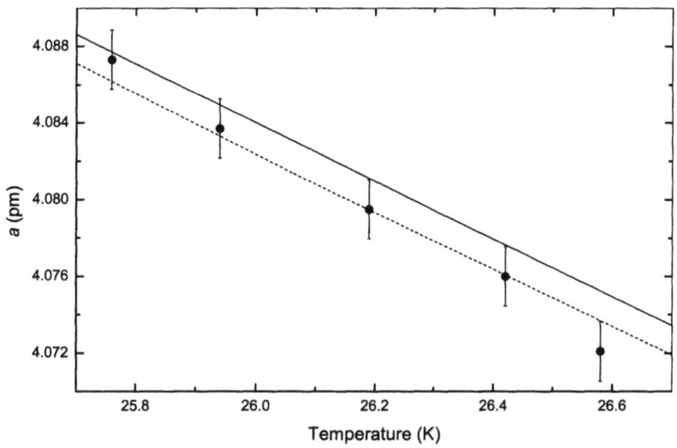

Fig. 23. Evidence of net vacancy content in fcc ^{4}He along the melting line, using (26). X–ray lattice parameters [18] are shown as (•), while the accepted (bulk) melting line is shown *solid.* X-ray data were taken with liquid and fcc solid coexisting in the cell.

including the need to work at high temperatures, one suspects that systematic errors influence the diverse reported results significantly.

For all three phases of ^{4}He, examples have been published: fcc [18], hcp [80], and bcc [81]. Figure 23 shows the results for fcc, obtained using synchrotron X–rays. The solid line represents standard bulk PVT equilibrium measurements. Because comparison not only depends upon accuracy of the individual measurements but also upon the temperature scale, it is fortunate that for highly compressible He another method is also applicable for T changes of a single He sample. In a rigid container, at constant macroscopic volume and constant number of atoms, (26) above simplifies for a cubic substance measured at fixed molar volume, V, to

$$c(T) = -3\frac{\Delta a(T)}{a_0} = \exp\left(-f(V,T)/k_B T\right) . \tag{27}$$

For non-cubic systems, it is necessary either to measure enough Bragg peaks to obtain the unit cell volume, or if that is not possible, to make assumptions about whether the cell changes shape during the processes of vacancy formation or annihilation as T is changed.

4.3 Inferred Activation Energies and Volumes for Vacancies

The direct diffraction method described in the previous section has produced some information on overall trends for vacancy free energies, f, in both bcc and hcp phases of both ^{3}He and ^{4}He [16]. There is considerable variation of f with molar volume, V. As expected, f decreases as the lattice is expanded, but there seems to be no great differences between values for different structures or for the two isotopes at the same molar volume. In ^{3}He, the trend of f versus V is smooth across the boundary between bcc (lower density) and hcp (higher

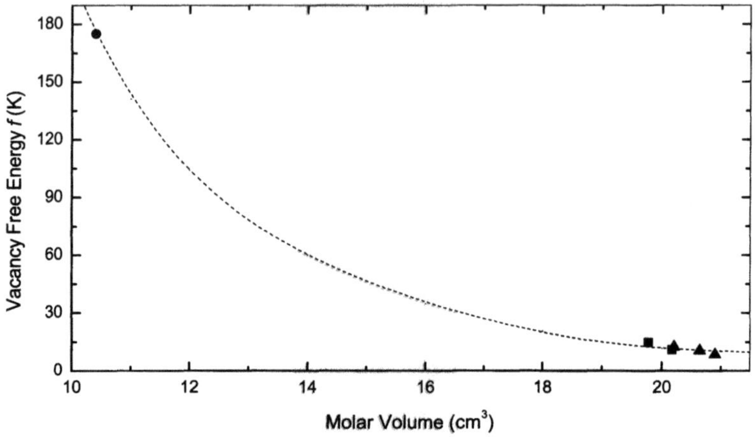

Fig. 24. Overall way that vacancy free energy in close–packed He solids depends upon the molar volume. Data are from [8,17,80]. The dotted line connecting the data follows the relation (24) of the text. From it one can obtain a value for the (V–dependent) free volume of formation of the vacancy. The full volume range shown in Fig. 2 is covered.

density) phases. This contrasts with a jump of the activation energy for self-diffusion as measured by NMR. From this, one infers that diffusion in hcp phase requires an activated jump much larger than that in bcc phase. Indeed, there is some evidence from different work that bcc phase diffusion involves tunnelling.

There is now available a rough idea of how this trend is extended over a very wide range of molar volumes of close–packed ^{4}He, through use of the data of Fig. 23 together with data at large molar volumes [17,80,82]. Use of the empirical relation (24) provides a way to bridge the wide current experimental gap in V, through use of the vacancy free volume, v, as essentially the only fitting parameter. The result is found in Fig. 24. Such a fitting scheme is a strong constraint upon viable vacancy models in close–packed He, provided there is no qualitative change in the vacancy microscopic structure over this range. On such an assumption, the result is that a value for the vacancy free volume near one–half the atomic volume is indicated.

Sophisticated computations on He defects are beginning to appear. In particular, the variational SWF method has produced some interesting results at large molar volume [14]. In hcp phase, a value of activation energy of 15.6 K was found. Note that the X-ray work measured concentrations, and hence found the vacancy free energy. Some room for discussion about the value of s still remains. In bcc phase an activation energy about 50% of that in hcp phase was found. The SWF work also studied relaxations of the atoms surrounding the vacancy site, and motions of the vacancies. Those in bcc phase were considerably more mobile. Computations over a broad range in V would be useful.

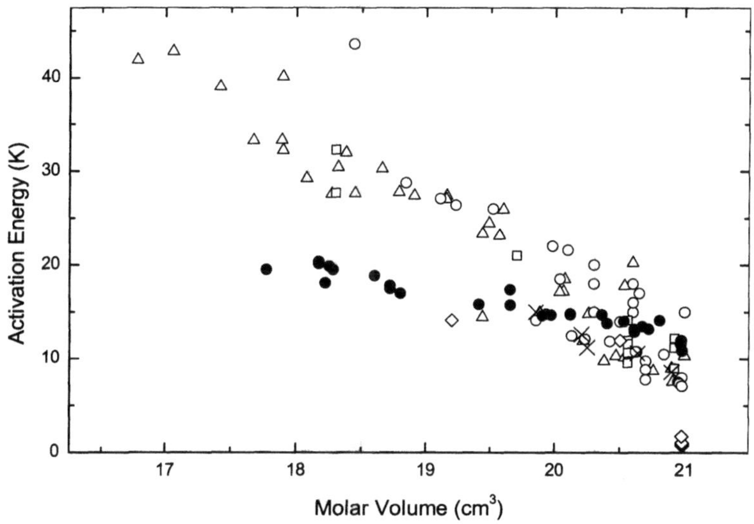

Fig. 25. From different experimental techniques, inferred values of activation energy in hcp ^{4}He have differed very widely. Some may apply to defect formation, some to defect diffusion, and some to other processes. ($\times$) X–ray data; ($\bullet$) pressure; ($\triangle$) NMR on included ^{3}He; ($\circ$) ion mobility; ($\square$) plastic deformation; ($\diamond$) ultrasonic attenuation.

4.4 Defects and Isotopic Mixtures Are a Zoological Garden

From different experimental techniques, inferred activation energies for vacancies in He have varied widely. Figure 25 shows examples of results for hcp ^{4}He [82]. These results show extraordinary differences and have led to wide variations in interpretation. These differences have both provoked much subsequent discussion and tended to damage confidence in any particular one of the experimental methods and its associated train of argument. Note again, the only absolute experimental method to obtain net vacancy content is by diffraction and by an associated bulk measurement. This method measures the net difference between atoms and substitutional atom sites, regardless of localization or association of the defects. At large V, the X–ray values lie near 12 K; they are the ones previously shown in Fig. 24. The X–ray method can contribute a useful constraint upon interpretation of other work, although it does not have the sensitivity to measure some small defect populations, nor the ability to discriminate among minority defect contributors (except in special cases like Kr where there is also evidence for interstitials [83]).

As a final example of experimental study of structural complexity in solid He, consider the equilibrium phase diagrams of solid mixtures of He isotopes. Near melting the two isotopes are miscible. At lower T's they exhibit isotopic phase separation. The particular T_{ps} depends upon the pressure and composition, and it is found that there are some deviations from ideal solution behavior. X–ray diffraction is a powerful method for investigating more complex phenomena in these mixtures, even though the X–ray form factor of each isotope is identical.

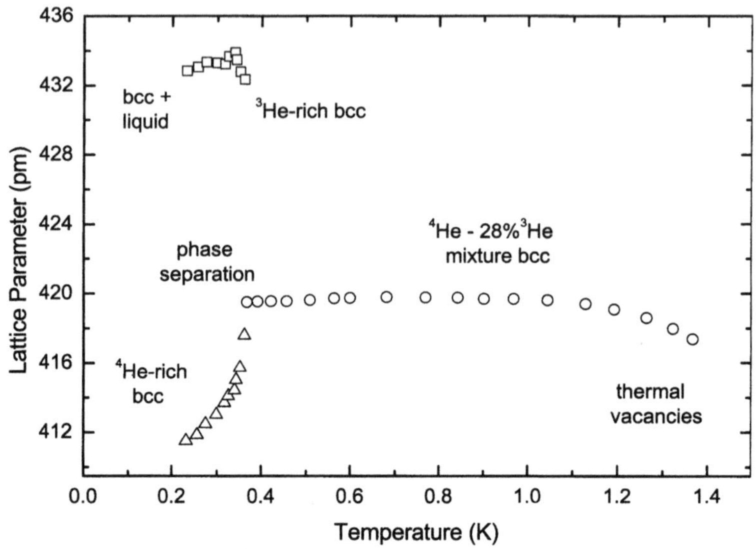

Fig. 26. Crystalline isotopic mixtures of ^{3}He and ^{4}He can be studied in bulk using X–ray scattering alone, because the lattice parameter is sensitive to the isotopic content [84]. Simultaneous phases can be identified, and extraordinary thermal vacancy content seen by the decrease in lattice parameter toward melting, according to (27).

The sample absorbs X–rays weakly, so that bulk measurements can be made in transmission geometry.

Results on a mixture containing 28% ^{3}He are shown in Fig. 26 [84]. For this homogeneous mixture at applied $P = 3.475$ MPa at a melting $T_{\mathrm{m}} = 1.58$ K, T_{ps} is sharply determined as 367±6 mK, where the (110) diffraction peak separates into two components upon cooling. The separation occurs rapidly owing to liquid–like diffusivity of the He isotopes in the solid phase. Because of the differing internal quantum pressures of each phase, they show different lattice parameters at the same (homogeneous) P. The larger lattice parameter is to be associated with the ^{3}He–rich phase; recall Fig. 4. In this case, parent and both children are bcc phase, a result directly given by the diffraction technique, and the crystal orientation of the children is within a few degrees of the single-crystal parent. When phase separation occurs, the lattice parameter differences between the two child phases is relatively large. It can therefore be concluded that there is atomic diffusion across local incoherent interfaces, because both diffraction peaks are seen everywhere in the sample using a scanned incident beam size of about 1 mm. In the figure, one sees that the ^{3}He–rich phase shows sudden compression upon cooling to 328 mK. This can be attributed to appearance of liquid phase, in analogy to melting which occurs in pure bcc ^{3}He upon cooling, at sufficiently low P. At high T can be seen the characteristic depression of the lattice parameter owing to thermal vacancy formation, (27). In this case, the magnitude of the net vacancy content is extraordinarily high, amounting to about 2% near melting. It can be speculated that the phenomenon is related to large effects seen in ordering alloys [77]. Further discussion appears in the original papers [84]. Diffraction data of this sort provide a rich vein for future investigations, especially if concurrent pressure measurements can be made.

5 Conclusion

This selective review has touched three areas in which diffraction and scattering experiments on condensed He phases can productively interact with detailed computations on these quantum systems over a wide range of experimental conditions:

- **Debye–Waller Effect** These traditional measurements, done at new levels of precision, motivate development of methods for extrapolating PIMC computations to the thermodynamic limit.
- **Kinetic energies, phonon excitations** High–energy pulsed neutrons and third–generation synchrotron X–rays provide new results in extended Brillouin zones, and for isotopic effects, to test microscopic models in detail.
- **Equilibrium defects** Direct X–ray data invite structural model–building over extended ranges of conditions (P, V, T).

I thank the organizers of the 257. Seminar in Bad Honnef for their invitation to participate and am indebted to the WE–Heraus–Stiftung for support to attend. The research of my group referred to herein was supported by the U.S. Department of Energy, Basic Energy Sciences, Division of Materials Sciences, under Contract DOE-DE-FG02-91ER45439.

References

1. R.O. Simmons: Z. Naturforsch. **48a**, 415 (1993).
2. R.O. Simmons: 'Neutron experiments on helium'. In: *Erforschung der Dynamik kondensierter Materie mit Synchrotronstrahlung*, ed. by E. Burkel, W. Schülke (Univ. Rostok, Rostock 1998), Vol II, pp. F4/1–33.
3. R.C. Blasdell, D.M. Ceperley, R.O. Simmons: Z. Naturforsch. **48a**, 433 (1993).
4. D.M. Ceperley, R.C. Blasdell, R.O. Simmons: Phys. Rev. Lett. **77**, 115 (1996).
5. K.A. Gernoth: 'The One- and Two-Body Densities of Crystalline Matter and Bragg and Diffuse Scattering of Neutrons and X-Rays'. In: *Particle Scattering, X–Ray Diffraction, and Microstructure of Solids and Liquids*, ed. by M. Ristig, K.A. Gernoth (Springer, Berlin 2002).
6. E. Burkel: *Inelastic Scattering of X–rays with Very High Energy Resolution*, Springer Tracts in Modern Physics, Vol.**125** (Springer, Berlin 1991); Rep. Prog. Phys. **63**, 171 (2000).
7. D.A. Arms: X–ray Debye–Waller Factor Measurements of Helium Crystals. Ph.D. thesis, University of Illinois at Urbana–Champaign (1999).
8. C.T. Venkataraman: X–ray Diffraction Study of the High Density Phases of Solid ^{4}He. Ph.D. thesis, University of Illinois at Urbana–Champaign (1996).
9. E. Draeger, D.M. Ceperley: Phys. Rev. **B 61**, 12094 (2000).

10. C. Seyfert: Untersuchung der Gitterdynamik von hcp Helium–3 und Helium–4 mit unelastischer Röntgenstreuung. Ph.D. thesis, University of Rostock (1998).
11. C. Seyfert, D.A. Arms, H. Sinn, R.O. Simmons, E. Burkel: Czech. J. Phys. **46**, 461 (1996); C. Seyfert, R.O. Simmons, H. Sinn, D.A. Arms, E. Burkel: J. Phys.: Condens. Matter **11**, 3501 (1999).
12. A.F. Andreev: 'Quantum Crystals'. In: *Progress in Low Temperature Physics*, ed. by D.F. Brewer (North–Holland, Amsterdam 1982) Vol. VIII, pp. 67–132. These ideas continue to stimulate experiments, e.g. V. Maidanov, A. Gaushin, V. Grigor'ev, A. Penzev, E. Rudavskii, A. Rybalko, Ye. Syrnikov: J. Low Temp. Phys. **126**, 133 (2002).
13. R.A. Guyer: 'Physics of Quantum Crystals'. In: *Solid State Physics*, ed. by F. Seitz, D. Turnbull, H. Ehrenreich (Academic Press, New York 1969) pp. 413–499.
14. B. Chauduri, F. Pederiva, G.V. Chester: Phys. Rev. **B 60**, 3271 (1999); J. Low Temp. Phys. **113**, 751 (1998) and references therein.
15. The early exception was J.H. Hetherington: Phys. Rev. **176**, 231 (1968); see also the later paper M.E.R. Bernier and J.H. Hetherington: Phys. Rev. **B 39**, 11285 (1989).
16. R.O. Simmons: J. Phys. Chem. Solids **55**, 85 (1994).
17. R.S. Shah: X–ray and Pressure Measurements on Helium Solids. Ph.D. thesis, University of Illinois at Urbana–Champaign (1999).
18. C.T. Venkataraman, R.O. Simmons: Czech. J. Physics **46**, 461 (1996).
19. A.Ya. Parshin: 'Crystallization Waves in Helium'. In: *Phonon Scattering in Condensed Matter*, ed. by W. Eisenmenger, K. Lassmann, S. Dottinger (Springer, Berlin 1984) pp. 226–33.
20. S.G. Lipson, E. Polturak: 'The Surface of Helium Crystals'. In: *Progress in Low Temperature Physics*, ed. by D.F. Brewer (Elsevier, Amsterdam 1987) Vol. XI, pp. 127-188; S. Balibar, P. Nozieres: Solid State Commun. **92**, 19 (1994).
21. S. Schöttl, K. Siemensmeyer, V. Boyko, I. Bat'ko, S. Mat'as, S. Raasch, E.D. Adams, T.E. Sherline: J. Low Temp. Phys. **126**, 51 (2002). In this case the formidable technical challenge is to keep the sample in an ordered magnetic state below 1 mK in spite of heating by neutron absorption.
22. D.A. Arms, R.O. Simmons, M. Schwoerer-Böhning, A.T. Macrander, T.J. Graber: Phys. Rev. Lett. **87**, 156402 (2001).
23. H.R. Glyde: *Excitations in Liquid and Solid Helium* (Clarendon Press, Oxford 1994)
24. R.A. Aziz, A.R. Janzen, R.O. Simmons: 'Rare Gases'. In: *Encyclopedia of Applied Physics*, ed. by G.L. Trigg (VCH Publishers, New York 1996) Vol. 16, pp. 71–96.
25. E.R. Dobbs: *Solid Helium Three* (Clarendon Press, Oxford 1994).
26. J. deBoer: Physica **14**, 139 (1948).
27. The author is indebted to M. Boninsegni and to D.M. Ceperley, respectively, for the Ne and He PIMC computations, which employed realistic pair potentials.
28. P. Schöffel, M. Müser: Phys. Rev. **B 63**, 224108 (2001).
29. T.R. Koehler: 'Lattice Dynamics of Quantum Crystals'. In: *Dynamical Properties of Solids*, ed. by G.K. Horton, A.A. Maradudin (American Elsevier, New York 1975) Vol. 2, pp. 1–104.
30. N.R. Werthamer: 'Self–Consistent Phonon Theory of Rare Gas Solids'. In: *Rare Gas Solids*, ed. by M.L. Klein, J.A. Venables (Academic Press, New York 1976) Vol. I, pp. 265–300.
31. D.M. Ceperley: Rev. Mod. Phys. **67**, 279 (1995).
32. F. Pederiva, S.A. Vitiello, K. Gernoth, S. Fantoni, L. Reatto: Phys. Rev. **B 53**, 15129 (1996).

33. H.H. Sample, C.A. Swenson: Phys. Rev. **158**, 188 (1967).
34. L.S. Salter: Adv. Phys. **14**, 1 (1965); T.H.K. Barron, G.K. White: *Heat Capacity and Thermal Expansion at Low Temperatures* (Kluwer Academic/Plenum, New York 1999).
35. D.C Wallace: *Thermodynamics of Crystals* (Dover Publications, New York 1998).
36. J.A. Leake, W.B. Daniels, J. Skalyo, Jr., B.C. Frazier, G. Shirane: Phys. Rev. **181**, 1251 (1969).
37. D. Acocella, G.K. Horton, E.R. Cowley: Phys. Rev. **B 61**, 8753 (2000).
38. S. Moroni, F. Pederiva, S. Fantoni, M. Boninsegni: Phys. Rev. Lett. **84**, 2650 (2000).
39. C.T. Venkataraman, D.A. Arms, R.S. Shah, R.O. Simmons: (to be published).
40. D.A. Arms, R.S. Shah, R.O. Simmons: (to be published).
41. A.T. Macrander, R.K. Crawford: phys. stat. sol. (a)**43**, 611 (1977); A.T. Macrander: Phys. Rev. **B 21**, 2549 (1980).
42. C.T. Venkataraman, R.O. Simmons: Rev. Sci. Instr. **67**, 3365 (1996) + CD–ROM.
43. G. Ahlers: Phys. Rev. **A 2**, 1505 (1970).
44. W.R. Gardner, J.K. Hoffer, N.E. Phillips: Phys. Rev. **A 7**, 1029 (1973).
45. A. Habenschuss, G.E. Ice, C.J. Sparks, R.A. Neiser: Nucl. Instrum. Methods **A 266**, 215 (1988).
46. J.C. Lang, G. Srajer, J. Wang, P.L. Lee: Rev. Sci. Instr. **70**, 4457 (1999).
47. C.A. Burns, E.D. Isaacs: Phys. Rev. **B 55** 5767 (1997).
48. C. Stassis, D. Khatamian, G.R. Kline: Solid State Commun. **25**, 531 (1978).
49. S.W. Lovesey: *Theory of Neutron Scattering from Condensed Matter* (Clarendon Press, Oxford 1984) Vol. 1: Nuclear Scattering.
50. G. Eckold: 'Inelastic Neutron Scattering from Structural Excitations'. In: *Particle Scattering, X–Ray Diffraction, and Microstructure of Solids and Liquids*, ed. by M. Ristig, K.A. Gernoth (Springer, Berlin 2002). Chap. 6
51. R. Senesi, C. Andreani, D. Colognesi, A. Cunsolo, M. Nardone: Phys. Rev. Lett. **86**, 4584 (2001); R. Senesi, C. Andreani, D. Colognesi: J. Low Temp. Phys. **126**, 57 (2002).
52. N. Schell, R.O. Simmons, E. Burkel: J. Synchrotron Rad. **3**, 316 (1996).
53. E. Burkel, C. Seyfert, Ch. Halcoussis, H. Sinn, R.O. Simmons: Physica **B 263–264**, 412 (1999)
54. J. Eckert, W. Thomlinson, G. Shirane: Phys. Rev. **B 16**, 1057 (1977); J. Eckert, W. Thomlinson, G. Shirane: Phys. Rev. **B 18**, 3074 (1978).
55. T.A. Kitchens, G. Shirane, V.J. Minkiewicz, E.B. Osgood: Phys. Rev. Lett. **29**, 552 (1972).
56. V.J. Minkiewicz, T.A. Kitchens, F.P. Lipschultz, R. Nathans, G. Shirane: Phys. Rev. **174**, 267 (1968); V.J. Minkiewicz, T.A. Kitchens, G. Shirane, E.B. Osgood: Phys. Rev. **A 8**, 1513 (1973).
57. W. Thomlinson, J. Eckert, G. Shirane: Phys. Rev. **B 18**, 1120 (1978).
58. R.A. Reese, S.K. Sinha, T.O. Brun, C.R. Tilford: Phys. Rev. **A 3**, 1688 (1971).
59. C.B. Ambegaokar, J.M. Conway, G. Baym: 'Inelastic Scattering of Neutrons by Anharmonic Crystals'. In: *Lattice Dynamics*, ed. by R.F. Wallis (Pergamon, Oxford 1965) pp. 261–270.
60. F.W. deWette, B.R.A. Nijboer: Phys. Lett. **18**, 19 (1965).
61. N.S. Gillis, T.R. Koehler, N.R. Werthamer: Phys. Rev. **175**, 1110 (1968).
62. G.L. Morley, K.L. Kliewer: Phys. Rev. **180**, 245 (1969).
63. W.M. Collins, H.R. Glyde: Phys. Rev. **B 18**, 1133 (1978).
64. R. Verbeni, A. Cunsolo, G. Pratesi, G. Monaco, F. Rosica, C. Masciovecchio, M. Nardone, G. Ruocco, F. Sette: Phys. Rev. **E 64**, 021203 (2001).

65. R.O. Simmons: 'Synchrotron radiation experiments on single crystals of helium'. In: *Erforschung der Dynamik kondensierter Materie mit Syncrhrotronstrahlung*, ed. by E. Burkel and W. Schülke (Univ. Rostok, Rostock 1998), Vol. II, pp. E6/1–20
66. C. Seyfert, R.O. Simmons, H. Sinn, D.A. Arms, E. Burkel:(to be published).
67. N.R. Werthamer: Solid State Commun. **9**, 2239 (1971).
68. In the case of H_2 crystals of similar size, the transformation from fcc to hcp and vice versa, with accompanying generation of stacking faults, was easily followed by X–rays. J.V. Gates, P.R. Granfors, B.A. Fraass, R.O. Simmons: Phys. Rev. **B 19**, 3667 (1979).
69. S.C. Lau, A.J. Dahm: J. Low Temp. Phys. **112**, 47 (1998).
70. I. Schuster, E. Polturak, Y. Swirsky, E.J. Schmidt, S.G. Lipson: J. Low Temp. Phys. **103**, 151 (1996); I. Berent, E. Polturak: J. Low Temp. Phys. **112**, 337 (1998); N. Gov, E. Polturak: Phys. Rev. **B 60**, 1019 (1999); N. Gov: Phys. Rev. **B 62**, 910 (2000); J. Low Temp. Phys. **121**, 731 (2000).
71. P. Remeijer, S.C. Steel, R. Jochemsen, G. Frossati: Low Temp. Phys. **23**, 438 (1997).
72. E.G. Kisvarsanyi, N.S. Sullivan: Phys. Rev. **B 55**, 3989 (1997); V.N. Grigor'ev: Phys. Rev. **B 55**, 3987 (1997); N.V. Zuev, V.V. Boiko, N.E. Dyumin, and V.N. Grigor'ev: J. Low Temp. Phys. **110**, 103 (1998); *ibid.* **111**, 597 (1998).
73. C.P. Flynn: *Point Defects and Diffusion* (Clarendon Press, Oxford 1972).
74. J. Holder, A.V. Granato: Phys. Rev. **182**, 729 (1969).
75. S.B. Trickey, W.P. Kirk, E.D. Adams: Rev. Mod. Phys. **44**, 668 (1982).
76. M. Roger, J.H. Hetherington, J.M. Delrieu: Rev. Mod. Phys. **55**, 1 (1983).
77. H.–E. Schaefer, K. Frenner, R. Würschum: Phys. Rev. **82**, 948 (1999).
78. K. Differt, A. Seeger, W. Trost: Mater. Sci. Forum **15–18**, 99 (1987), key data from W. Trost Ph.D. thesis, University of Stuttgart (1986).
79. R.O. Simmons, R.W. Balluffi: Phys. Rev. **129**, 1533 (1963).
80. B.A. Fraass, P.R. Granfors, R.O. Simmons: Phys. Rev. **B 39**, 124 (1989).
81. P.R. Granfors, B.A. Fraass, and R.O. Simmons: J. Low Temp. Phys. **67**, 353 (1987).
82. R.S. Shah, C.T. Venkataraman, R.O. Simmons: (to be published).
83. C.A. Gordon, A.V. Granato, R.O. Simmons: J. Non–Cryst. Solids **205–207**, 216 (1996).
84. B.A. Fraass, R.O. Simmons: Phys. Rev. **B 36**, 97 (1987); Phys. Rev. **B 37**, 5058 (1988). See also S.N. Ehrlich, R.O. Simmons: J. Low Temp. Phys. **68**, 125 (1987).

The One- and Two-Body Densities of Crystalline Matter and Bragg and Diffuse Scattering of Neutrons and X-Rays

Klaus A. Gernoth

Department of Physics, UMIST, P.O. Box 88, Manchester M60 1QD, United Kingdom

Abstract. This work presents a theoretical treatment of the Bragg and diffuse scattering of an incident double-beam of neutrons or X-rays from crystalline matter. The incoming double-beam is a linear superposition of two plane waves with the same energy. It is shown that the total Bragg plus diffuse scattering differential cross section is a sum of terms stemming from the scattering of the two individual plane waves in the incoming double-beam and of terms originating in the interference of the scattering of the two superimposed incident plane waves. The leading-order dominant Bragg peaks in the total differential cross section are size-extensive and depend on the lattice Fourier transform of the one-body density of the scattering material. It is proven that the Bragg peaks are a consequence of the long-range behavior of the two-body density for large interparticle distances. The diffuse background scattering contributions in the total differential cross section for the scattering of a double-beam depend on the lattice Fourier transform of the one-body density and, more importantly, on the continuous Fourier transform of the (discrete) lattice Fourier transform of the two-body density of the scatterer, which latter transform is a function of the relative position vector of the two particles. One of the most significant conclusions from these findings is that the *full* two-body density of a crystalline material is experimentally accessible by means of the diffuse background scattering of a double-beam of neutrons or X-rays in which the difference of the two wave number vectors in the incoming two-beam is a reciprocal lattice vector of the scattering crystal. This paper deals also with the crystallographic symmetries in the Bragg and the diffuse scattering structure functions. Based on results of previous research, it is found that both scattering structure functions follow exactly the same crystallographic symmetry patterns as the lattice Fourier transform functions of the two-body density do. The formal theory of Bragg and diffuse scattering of a neutron or X-ray two-beam derived here may be readily related to previous and ongoing exact numerical Monte Carlo calculations of the spatial microstructure of crystals, in which the local one- and two-body densities are computed by means of exact group-theoretical Fourier path integral Monte Carlo simulations.

1 Introduction

In a series of recent theoretical papers [1–6] the crystallographic point and space group symmetries of the local one- and two-body densities of crystals have been thoroughly analyzed both analytically [1–4,6] and numerically in specific Fourier path integral Monte Carlo (FPIMC) applications to rare gas crystals in the hexagonal close-packed (hcp) structure [3,4,6]. The general analytical group-theoretical findings reported and mathematically rigorously proven in these publications [1–4] hold for any crystallographic space group. In conjunction with exact computational techniques, such as Monte Carlo (MC) or molecular dynamics (MD) simulations, these group-theoretical methods allow for exact numerical calculations of the local one- and two-body densities of crystalline matter. This is demonstrated in Refs. 1, 3, 4, and 6, in which the general analytical results derived for the one- and two-body densities within a framework of crystallographic point and space group theory are specialized to crystallographic space group $P6_3/mmc$ (D_h^6 in Schoenflies notation), which is the space group of the hcp structure. In specific numerical FPIMC applications [3,4,6] to rare gas crystals in the hcp solid phase numerical results for the local one- and two-body densities have been computed along with the macroscopic thermodynamic kinetic, potential, and total energies. Although these macroscopic quantities are of theoretical and experimental interest in their own right, the emphasis in these works [1–6] clearly is on the spatial microstructure of the inert gas crystals as embodied in the crystallographic point and space group symmetries of the one- and two-body densities.

The group-theoretical methods developed and tested numerically in exact FPIMC simulations of noble gas crystals [3,4,6] lend themselves also to carrying out a complete classification for the liquid-solid phase transition of symmetry-breaking terms in the one-and two-body densities according to irreducible order parameter symmetries of the space group of the high-symmetry liquid phase [2,4], which is the extended Euclidean space group $E(|R^3)$ in three dimensions of all rotations and translations and all combinations thereof. This classification was undertaken in full in Ref. 4 for the solidification of a spatially homogeneous uniform liquid into a crystal with crystallographic space group $P6_3/mmc$. The same work [4] reports and discusses in detail the numerical results for the symmetry-breaking terms in the one- and two-body densities of an exact FPIMC simulation of the spontaneous crystallization of inert gas liquids into solid hcp phases, the crystallographic space group of which is $P6_3/mmc$. One of the important findings of the group-theoretical analysis of spontaneous solidification in the one- and two-body densities performed in Ref. 4 is that the vast majority of symmetry-breaking irreducible order parameter symmetries does not appear as symmetry-breaking terms in the one-body density but only in the symmetry-breaking terms of the two-body density, when the system undergoes the liquid-solid phase transition. As proven in Refs. 2 and 4 this behavior of the one- and two-body densities in the liquid-solid phase transition is due to the fact that the vast majority of irreducible representations of the high-symmetry space group $E(|R^3)$ of the uniform liquid does not possess basis functions in the space of functions of a single three-dimensional variable but only in the space of functions of two three-dimensional variables. The reader is referred to Ref. 4 for a thorough treatment of symmetry-breaking in the one- and two-body densities upon solidification and a discussion of its novelty in the context of the Landau theory of phase transitions [7,8].

Although the specific applications of the crystallographic point and space group theory of the one- and two-body densities as derived in Refs. 1–6 are to noble gas condensed systems at finite temperatures, this general theory is not limited in its applicability to FPIMC simulations of crystals and of spontaneous crystallization but is of general validity and may be employed within any theoretical model of condensed matter systems that deals with the spatial microstructure of materials in terms of the local one- and two-body densities. This applies in particular to correlated many-body wave functions [1], where incipient successes have been achieved in correlated wave functions treatments of solidification [9,10], and to shadow wave functions Monte Carlo simulations [1,11] of the helium liquids and solids. The group-theoretical analysis of symmetries and symmetry-breaking in the one- and two-body densities may be employed also in exact Green function and diffusion Monte Carlo calculations of the spatial microstructure of correlated quantum many-body systems at vanishing temperature. Beyond that, the successful merger of crystallographic point and space group theory with microscopic quantum many-body theory achieved in Refs. 1–6 has proven useful also in the coupled-cluster theory of spin lattice systems [13–15].

References 3 and 4 address also the relationship of theoretical results computed within the exact group-theoretical FPIMC approach developed and numerically tested in these works [3,4,6] to experimentally measurable quantities. We mention in particular that the kinetic energy per particle, one of *the* macroscopic thermodynamic quantities that may be computed theoretically by means of exact Monte Carlo integration techniques, may be obtained from the double differential cross section measured in deep-inelastic scattering of neutrons from intense spallation sources [3,4,16–19]. Furthermore, the magnitude $|\rho(\mathbf{K})|$ of the Fourier transform $\rho(\mathbf{K})$ of the one-body density $\varrho(\mathbf{x}) = \sum_{\mathbf{K}} \rho(\mathbf{K}) \mathrm{e}^{i\,\mathbf{K}\mathbf{x}}$ (series over all reciprocal lattice vectors $\mathbf{K}$) may be obtained from the intensities of the Bragg scattering peaks in neutron scattering and X-ray diffraction experiments [3,4]. For a crystal the two-body density $\varrho_2(\mathbf{S}, \mathbf{r})$may be written in the form of a reciprocal lattice Fourier series in the center-of-mass variable $\mathbf{S}$ of the two particles [1–4,6], i.e. $\varrho_2(\mathbf{S}, \mathbf{r}) = \sum_{\mathbf{K}} u(\mathbf{r}; \mathbf{K}) \mathrm{e}^{i\,\mathbf{K}\mathbf{S}}$ (series over all reciprocal lattice vectors $\mathbf{K}$) with a lattice Fourier transform $u(\mathbf{r}; \mathbf{K})$, that is a function of the relative position vector $\mathbf{r}$ of the two particles. It turns out [3,4] that the lowest-lying $\mathbf{K} = 0$ Fourier coefficient function $u(\mathbf{r}; 0)$ in such an expansion of the two-body density $\varrho_2(\mathbf{S}, \mathbf{r})$ may be obtained from the diffuse background scattering in neutron scattering and X-ray diffraction experiments with an incident single-beam of the form $\mathrm{e}^{i\,\mathbf{k}_{\mathrm{in}}\mathbf{x}}$, where $\mathbf{k}_{\mathrm{in}}$ is the wave number vector of the incoming beam.

That it is possible to obtain experimental results for $u(\mathbf{r}; 0)$ in the lattice Fourier series representation $\varrho_2(\mathbf{S}, \mathbf{r}) = \sum_{\mathbf{K}} u(\mathbf{r}; \mathbf{K}) \mathrm{e}^{i\,\mathbf{K}\mathbf{S}}$ of the full two-body density $\varrho_2(\mathbf{S}, \mathbf{r})$ from the diffuse background scattering of a single-beam of neutrons or X-rays raises the question, which kind of experiment would yield experimental access to the higher-lying $\mathbf{K} \neq 0$ Fourier coefficient functions $u(\mathbf{r}; \mathbf{K})$ of the two-body density $\varrho_2(\mathbf{S}, \mathbf{r})$ and thus would allow to map out the full two-body density $\varrho_2(\mathbf{S}, \mathbf{r})$ of a material rather than just the $\mathbf{K} = 0$ term $u(\mathbf{r}; 0)$.

In this work it will be shown that the $\mathbf{K} \neq 0$ component functions $u(\mathbf{r}; \mathbf{K})$ of the two-body density may be related to the diffuse background scattering of an incident double-beam of neutrons or X-rays. The incoming double-beam has to be of the form $e^{i\mathbf{k}_1^{(\mathrm{in})}\mathbf{x}} + Ce^{i\mathbf{k}_2^{(\mathrm{in})}\mathbf{x}}$ (C an in general complex-valued constant) of a linear superposition of two plane waves $e^{i\mathbf{k}_1^{(\mathrm{in})}\mathbf{x}}$ and $e^{i\mathbf{k}_2^{(\mathrm{in})}\mathbf{x}}$ with the same wave number, i.e. $\left|\mathbf{k}_1^{(\mathrm{in})}\right| = \left|\mathbf{k}_2^{(\mathrm{in})}\right|$, and the difference $\mathbf{k}_1^{(\mathrm{in})} - \mathbf{k}_2^{(\mathrm{in})}$ must be a reciprocal lattice vector $\mathbf{K} = \mathbf{k}_1^{(\mathrm{in})} - \mathbf{k}_2^{(\mathrm{in})}$ of the scattering material, in order to obtain $u(\mathbf{r};\mathbf{K})$ from the diffuse background scattering in such a two-beam scattering experiment.

In Sect. 2 of this paper we will derive an expression for the total (Bragg plus diffuse) differential cross section in the Born approximation for the scattering of a double-beam of neutrons of the form explained above in the preceding paragraph. We follow closely the derivation of the single-beam differential cross section in the Born approximation that is given in Ref. 20 for the scattering of neutrons in conventional single-beam experiments and generalize it to the case of scattering of an incoming double-beam of neutrons. The formula derived in Sect. 2 for the two-beam differential cross section holds for nearly elastic scattering of neutrons in the so-called static approximation [3,4,20].

In Sect. 3 we reformulate the total (Bragg plus diffuse) double-beam scattering structure function in terms of the one- and two-body densities $\varrho(\mathbf{x})$ and $\varrho_2(\mathbf{S},\mathbf{r})$ of the scatterer. It is shown that the leading-order dominant Bragg peaks in the total double-beam scattering structure function originate in the asymptotic long-range behavior of the two-body density and involve only the lattice Fourier transform $\rho(\mathbf{K})$ of the one-body density $\varrho(\mathbf{x})$. The diffuse scattering structure function in the total two-beam scattering structure function may be expressed in terms of $\rho(\mathbf{K})$ and in terms of the continuous Fourier transforms of the $\mathbf{K}$-components $u(\mathbf{r};\mathbf{K})$ of the two-body density $\varrho_2(\mathbf{S},\mathbf{r})$. The double-beam scattering structure function reduces to the usual single-beam scattering result for an incident single-beam. The limit of double-beam scattering from a spatially homogeneous quantum many-body system, such as a uniform liquid, is briefly discussed. The total double-beam scattering structure function splits up into contributions from the scattering of the individual beams $e^{i\mathbf{k}_1^{(\mathrm{in})}\mathbf{x}}$ and $e^{i\mathbf{k}_2^{(\mathrm{in})}\mathbf{x}}$ in the incident superimposed two-beam $e^{i\mathbf{k}_1^{(\mathrm{in})}\mathbf{x}} + Ce^{i\mathbf{k}_2^{(\mathrm{in})}\mathbf{x}}$ and into contributions from the interference of the scattering of the two incoming superimposed planar waves.

Section 4 furnishes a complete analysis of the crystallographic symmetries in the Bragg scattering and diffuse scattering structure functions. The formal results presented in Sect. 4 for the double-beam scattering structure functions are deduced with the help of the formal crystallographic point and space group theory of the one- and two-body densities as derived in Refs. 1–6. The reader is referred in particular to Refs. 1–4 for a much more comprehensive and detailed presentation of crystallographic point and space group symmetries in the one- and two-body densities than is given here in Sect. 4. Excellent accounts of crystallographic point and space groups and of their irreducible representations may be found in Refs. 21 and 22. The final conclusion of Sect. 4 is that the Bragg and diffuse double-beam scattering structure functions follow exactly the same crystallographic symmetries as the **K**-components $u(\mathbf{r};\mathbf{K})$ in the two-body density $\varrho_2(\mathbf{S},\mathbf{r})$ do.

Section 5 is devoted to relating the Bragg and diffuse double-beam scattering structure functions, defined in Sect. 3 in terms of $\rho(\mathbf{K})$ and $u(\mathbf{r};\mathbf{K})$, to the total (Bragg plus diffuse) differential cross section of the diffraction of a double-beam of X-rays. Since in contrast to neutrons, which are scattered by the nuclei of the target system, X-rays are scattered by the electrons of the scattering many-body system, the relation derived in Sect. 5 for the double-beam X-ray differential cross section involves the atomic form factor, which is nothing else but the Fourier transform of the distribution of the shell electrons of an atom of the scatterer about the nucleus in the center of the atom. Since the atomic form factors are known and may be taken from the International Tables for Crystallography (ITC), Vol. C [23], the relation for the two-beam X-ray differential cross section allows experimental access to the lattice Fourier transform $\rho(\mathbf{K})$ of the atomic one-body density $\varrho(\mathbf{x})$ of the scattering many-body system and to the lattice Fourier transform functions $u(\mathbf{r};\mathbf{K})$ of the atomic two-body density $\varrho_2(\mathbf{S},\mathbf{r})$. Also the diffraction of an incoming X-ray double-beam of the form $e^{i\,\mathbf{k}_1^{(\mathrm{in})}\mathbf{x}} + Ce^{i\,\mathbf{k}_2^{(\mathrm{in})}\mathbf{x}}$ requires the difference $\mathbf{K} = \mathbf{k}_1^{(\mathrm{in})} - \mathbf{k}_2^{(\mathrm{in})}$ to be a reciprocal lattice vector of the scatterer, in order to produce experimental information about $u(\mathbf{r};\mathbf{K})$. As for double-beam neutron scattering also the double-beam X-ray diffraction differential cross section consists of dominant leading-order Bragg peaks, depending only on $\rho(\mathbf{K})$ and the atomic form factor, and of diffuse background scattering terms, depending on the atomic form factor and on both $\rho(\mathbf{K})$ and the continuous Fourier transform of the **K**-component $u(\mathbf{r};\mathbf{K})$ of the atomic two-body density $\varrho_2(\mathbf{S},\mathbf{r})$ of the scatterer. As for neutrons also the two-beam X-ray diffraction differential cross section contains contributions from the scattering of the individual beams $e^{i\,\mathbf{k}_1^{(\mathrm{in})}\mathbf{x}}$ and $e^{i\,\mathbf{k}_2^{(\mathrm{in})}\mathbf{x}}$ in the incident double-beam $e^{i\,\mathbf{k}_2^{(\mathrm{in})}\mathbf{x}} + Ce^{i\,\mathbf{k}_2^{(\mathrm{in})}\mathbf{x}}$ and contributions from the interference of the diffraction of the two incoming superimposed plane waves.

2 Theory of Bragg and Diffuse Scattering of a Double-Beam of Neutrons from Condensed Matter

In this section we will derive a formula for the differential cross section $\frac{d\sigma}{d\Omega}$ for the scattering of a double-beam of neutrons, that in the following section will allow us to relate the local one- and two-body densities of the scattering many-body system to such two-beam experiments.

To this end we commence with the stationary Schrödinger equation of the total system, consisting of the particle and the scatterer,

$$\left\{\hat{H}_0 - \frac{\hbar^2}{2m_\mathrm{n}}\nabla^2 + V(\mathbf{x},\mathbf{R})\right\}\Phi(\mathbf{x},\mathbf{R}) = E\Phi(\mathbf{x},\mathbf{R})\,. \tag{1}$$

Here $\hat{H}_0$ is the Hamiltonian of the scattering many-body system alone and E the energy eigenvalue of the total system. The second term in the operator on the left side of (1) is the operator of the kinetic energy of the neutron of mass m_{n}, acting on $\mathbf{x} \in |\mathrm{R}^3$, the neutron's position in three-dimensional space $|\mathrm{R}^3$. The interaction of the neutron with the scatterer is denoted by $V(\mathbf{x}, \mathbf{R})$, where the $3N$-dimensional vector $\mathbf{R} = (\mathbf{r}_1, \mathbf{r}_2, \cdots, \mathbf{r}_N)$ represents the Cartesian coordinates of the N particles of the scattering many-body system. The three Cartesian coordinates of a particle i, with $1 \leq i \leq N$, of the scatterer are denoted by the three-dimensional vector $\mathbf{r}_i$.

The stationary wave function $\Phi(\mathbf{x}, \mathbf{R})$ of the total system may be expanded in terms of a complete orthonormalized set of eigenfunctions $\Psi_\lambda(\mathbf{R})$ of $\hat{H}_0$, given by the stationary Schrödinger equation

$$\hat{H}_0 \Psi_\lambda(\mathbf{R}) = E_\lambda \Psi_\lambda(\mathbf{R}) \tag{2}$$

of the scattering system alone. The subscript λ denotes a complete set of quantum numbers characterizing the many-body eigenfunctions $\Psi_\lambda(\mathbf{R})$ of $\hat{H}_0$ and the associated energy eigenvalues E_λ. The expansion of $\Phi(\mathbf{x}, \mathbf{R})$ in terms of the wave functions $\Psi_\lambda(\mathbf{R})$ may be cast in the form

$$\Phi(\mathbf{x}, \mathbf{R}) = \sum_\lambda A_\lambda(\mathbf{x}) \Psi_\lambda(\mathbf{R}), \tag{3}$$

wherein the series with $\mathbf{x}$-dependent expansion coefficients $A_\lambda(\mathbf{x})$ extends over the complete set of eigenfunctions $\Psi_\lambda(\mathbf{R})$ of $\hat{H}_0$.

Inserting the expansion (3) in the Schrödinger equation (1), carrying out a scalar multiplication with $\Psi_{\lambda'}(\mathbf{R})$, and exploiting the orthonormality relations

$$\int \Psi_{\lambda'}^*(\mathbf{R}) \Psi_\lambda(\mathbf{R})\, d\mathbf{R} = \int \Psi_{\lambda'}^*(\mathbf{R}) \Psi_\lambda(\mathbf{R})\, d\mathbf{r}_1 \cdots d\mathbf{r}_N = \delta_{\lambda\lambda'} \tag{4}$$

obeyed by the orthonormalized eigenfunctions $\Psi_\lambda(\mathbf{R})$ of $\hat{H}_0$ yields the inhomogeneous differential equation

$$\left\{\nabla^2 + k_{\lambda'}^2\right\} A_{\lambda'}(\mathbf{x}) = \frac{2m_{\mathrm{n}}}{\hbar^2} \int \Psi_{\lambda'}^*(\mathbf{R}) V(\mathbf{x}, \mathbf{R}) \Phi(\mathbf{x}, \mathbf{R})\, d\mathbf{R} \tag{5}$$

for $A_{\lambda'}(\mathbf{x})$, where for every particle i of the scatterer the three-dimensional integration $\int \cdots d\mathbf{r}_i$ in the $3N$-dimensional integral $\int \cdots d\mathbf{R}$ in the source term on the right side of (5) extends over the volume V that is occupied by the scattering system. Here we have introduced

$$k_{\lambda'} = \sqrt{\frac{2m_{\mathrm{n}}}{\hbar^2} (E - E_{\lambda'})}\,. \tag{6}$$

The asterisk denotes complex conjugation and $\delta_{\lambda\lambda'}$ in (4) is the Kronecker delta, $\delta_{\lambda\lambda'} = 0$ for $\lambda \neq \lambda'$ and $\delta_{\lambda\lambda'} = 1$ for $\lambda = \lambda'$.

The to (5) corresponding homogeneous differential equation reads

$$\left\{\nabla^2 + k_{\lambda'}^2\right\} A_{\lambda'}^{(\mathrm{hom})}(\mathbf{x}) = 0\,. \tag{7}$$

The solutions of (5) for $A_{\lambda'}(\mathbf{x})$ may be constructed from the solutions of (7) by adding the convolution of the Green function $G_{\lambda'}(\mathbf{x}-\mathbf{x}')$ of the operator $\nabla^2 + k_{\lambda'}^2$ with the inhomogeneous source term of the right of (5). The Green function $G_{\lambda'}(\mathbf{x}-\mathbf{x}')$ in turn is given by the solution of the inhomogeneous differential equation

$$\left\{\nabla^2 + k_{\lambda'}^2\right\} G_{\lambda'}(\mathbf{x}-\mathbf{x}') = \delta(\mathbf{x}-\mathbf{x}') \tag{8}$$

with the Dirac δ-distribution $\delta(\mathbf{x}-\mathbf{x}')$ in three dimensions as source term. The Green function $G_{\lambda'}(\mathbf{x}-\mathbf{x}')$ is the kernel of the resolvent (integral) operator of the (differential) operator $\nabla^2 + k_{\lambda'}^2$ and in the present case assumes the shape

$$G_{\lambda'}(\mathbf{x}-\mathbf{x}') = -\frac{1}{4\pi} \frac{\mathrm{e}^{i\,k_{\lambda'}|\mathbf{x}-\mathbf{x}'|}}{|\mathbf{x}-\mathbf{x}'|}\,. \tag{9}$$

The δ-distribution $\delta(\mathbf{x}-\mathbf{x}')$ in (8) constitutes in a similar vein the kernel of the identity (integral) operator in Hilbert space. The solutions of (7) for $A_{\lambda'}^{(\mathrm{hom})}(\mathbf{x})$ correspond to the asymptotic situation where the neutron is far away from the scatterer, i.e. $|\mathbf{x}-\mathbf{r}_i| \longrightarrow \infty$ for all particles $1 \le i \le N$ in the scatterer, so that the neutron-scatterer interaction $V(\mathbf{x},\mathbf{R})$ vanishes in this asymptotic regime and, accordingly, the expression on the right side of (5) becomes zero, yielding (7).

We now consider solutions of the homogeneous differential equation (7) for the $\mathbf{x}$-dependent expansion coefficients $A_{\lambda'}^{(\mathrm{hom})}(\mathbf{x})$ of the form

$$A_{\lambda'}^{(\mathrm{hom})}(\mathbf{x}) = \left\{\mathrm{e}^{i\,\mathbf{k}_\lambda^{(1)}\mathbf{x}} + C\mathrm{e}^{i\,\mathbf{k}_\lambda^{(2)}\mathbf{x}}\right\}\delta_{\lambda\lambda'}$$
$$\text{with} \quad k_\lambda = k_\lambda^{(1)} = \left|\mathbf{k}_\lambda^{(1)}\right| = k_\lambda^{(2)} = \left|\mathbf{k}_\lambda^{(2)}\right|\,, \tag{10}$$

where C may be any (constant) complex number. Obviously, $A_{\lambda'}^{(\mathrm{hom})}(\mathbf{x}) = 0$ for $\lambda' \neq \lambda$ in (10) and for $\lambda' = \lambda$ in general is the superposition of two plane waves with the same wave number k_λ. By making use of (8) for the Green function $G_{\lambda'}(\mathbf{x}-\mathbf{x}')$ it then is fairly straightforward to prove that

$$A_{\lambda'}(\mathbf{x}) = \left\{\mathrm{e}^{i\,\mathbf{k}_\lambda^{(1)}\mathbf{x}} + C\mathrm{e}^{i\,\mathbf{k}_\lambda^{(2)}\mathbf{x}}\right\}\delta_{\lambda\lambda'}$$
$$+\frac{2m_\mathrm{n}}{\hbar^2}\int G_{\lambda'}(\mathbf{x}-\mathbf{x}')\Psi_{\lambda'}^*(\mathbf{R})V(\mathbf{x}',\mathbf{R})\Phi(\mathbf{x}',\mathbf{R})\,d\mathbf{x}'d\mathbf{R} \tag{11}$$

are solutions of the inhomogeneous differential equation (5) for the expansion coefficients $A_{\lambda'}(\mathbf{x})$ in the expression (3) for the solution $\Phi(\mathbf{x},\mathbf{R})$ of the neutron-plus-scatterer Schrödinger equation (1).

Inserting $A_{\lambda'}(\mathbf{x})$ from (11) in the expansion (3) for $\Phi(\mathbf{x},\mathbf{R})$ yields the result

$$\Phi(\mathbf{x},\mathbf{R}) = \left\{\mathrm{e}^{i\,\mathbf{k}_\lambda^{(1)}\mathbf{x}} + C\mathrm{e}^{i\,\mathbf{k}_\lambda^{(2)}\mathbf{x}}\right\}\Psi_\lambda(\mathbf{R})$$
$$+\frac{2m_\mathrm{n}}{\hbar^2}\sum_{\lambda'}\left\{\int G_{\lambda'}(\mathbf{x}-\mathbf{x}')\Psi_{\lambda'}^*(\mathbf{R}')V(\mathbf{x}',\mathbf{R}')\Phi(\mathbf{x}',\mathbf{R}')\,d\mathbf{x}'d\mathbf{R}'\right\}$$
$$\times\,\Psi_{\lambda'}(\mathbf{R})\,. \tag{12}$$

The first term in this formula for $\Phi(\mathbf{x},\mathbf{R})$ is a solution of the stationary Schrödinger equation for the neutron-plus-scatterer system in the absence of an interaction between the neutron and the scattering many-body system, i.e. a solution of (1) for $V(\mathbf{x},\mathbf{R})=0$. The other terms in (12) are generated by the neutron-scatterer interaction $V(\mathbf{x},\mathbf{R})$.

To proceed from there, we next consider the shape of the wave function $\Phi(\mathbf{x},\mathbf{R})$ in the asymptotic region where the neutron is far away from the scatterer, i.e. for $|\mathbf{x}-\mathbf{r}_i|\gg 1$ for all particles $1\le i\le N$ of the scatterer. The multi-dimensional integrals in (12) yield a contribution only for $\mathbf{x}'$ in or close to the volume occupied by the scatterer, because the interaction $V(\mathbf{x}',\mathbf{R}')$ vanishes for $\mathbf{x}'$ far away from the scattering system. This implies $|\mathbf{x}'|\ll|\mathbf{x}|=x$ and whence

$$|\mathbf{x}-\mathbf{x}'|\approx x-\frac{\mathbf{x}\mathbf{x}'}{x} \qquad \text{for } |\mathbf{x}'|\ll|\mathbf{x}| \tag{13}$$

in the asymptotic regime where the neutron is far away from the scatterer. Using then the approximation

$$\frac{\mathrm{e}^{i\,k_{\lambda'}|\mathbf{x}-\mathbf{x}'|}}{|\mathbf{x}-\mathbf{x}'|}\approx\frac{\mathrm{e}^{i\,k_{\lambda'}x}}{x}\,\mathrm{e}^{-i\,k_{\lambda'}(\hat{\mathbf{x}}\mathbf{x}')} \qquad \text{for } |\mathbf{x}'|\ll|\mathbf{x}|\,, \tag{14}$$

where $\hat{\mathbf{x}}=\mathbf{x}/x$ is the unit vector in the direction of $\mathbf{x}$, in the Green function (9) the asymptotic behavior $\Phi_{\mathrm{as}}(\mathbf{x},\mathbf{R})$ of $\Phi(\mathbf{x},\mathbf{R})$ of (12) may be cast in the form

$$\begin{aligned}\Phi_{\mathrm{as}}(\mathbf{x},\mathbf{R}) = {} & \left\{\mathrm{e}^{i\,k_\lambda\left(\hat{\mathbf{k}}_1^{(\mathrm{in})}\mathbf{x}\right)}+C\mathrm{e}^{i\,k_\lambda\left(\hat{\mathbf{k}}_2^{(\mathrm{in})}\mathbf{x}\right)}\right\}\Psi_\lambda(\mathbf{R}) \\ & +\sum_{\lambda'} f_{\lambda\lambda'}\left(\mathbf{k}_1^{(\mathrm{in})},\mathbf{k}_2^{(\mathrm{in})},\mathbf{k}';C\right)\frac{\mathrm{e}^{i\,k_{\lambda'}x}}{x}\,\Psi_{\lambda'}(\mathbf{R})\,,\end{aligned} \tag{15}$$

wherein $\mathbf{k}_1^{(\mathrm{in})}=\mathbf{k}_\lambda^{(1)}$, $\mathbf{k}_2^{(\mathrm{in})}=\mathbf{k}_\lambda^{(2)}$, $\mathbf{k}'=k_{\lambda'}\hat{\mathbf{x}}$,

$$f_{\lambda\lambda'}\left(\mathbf{k}_1^{(\mathrm{in})},\mathbf{k}_2^{(\mathrm{in})},\mathbf{k}';C\right)=-\frac{m_{\mathrm{n}}}{2\pi\hbar^2}\int\mathrm{e}^{-i\,\mathbf{k}'\mathbf{x}'}\Psi^*_{\lambda'}(\mathbf{R}')V(\mathbf{x}',\mathbf{R}')\Phi(\mathbf{x}',\mathbf{R}')\,d\mathbf{x}'d\mathbf{R}'\,, \tag{16}$$

and where $\hat{\mathbf{k}}_1^{(\mathrm{in})}=\mathbf{k}_\lambda^{(1)}/k_\lambda=\mathbf{k}_1^{(\mathrm{in})}/k_\lambda$ and $\hat{\mathbf{k}}_2^{(\mathrm{in})}=\mathbf{k}_\lambda^{(2)}/k_\lambda=\mathbf{k}_2^{(\mathrm{in})}/k_\lambda$ are the unit vectors in the directions of the two superimposed incoming plane waves in (10).

The physical meaning of the asymptotic form (15) of the solution $\Phi(\mathbf{x},\mathbf{R})$ of the Schrödinger equation (1) is clear now. The scatterer is in the initial eigenstate $\Psi_\lambda(\mathbf{R})$ with energy eigenvalue E_λ and is hit by an incoming neutron beam that is given by the superposition $\mathrm{e}^{i\,k_\lambda\left(\hat{\mathbf{k}}_1^{(\mathrm{in})}\mathbf{x}\right)}+C\mathrm{e}^{i\,k_\lambda\left(\hat{\mathbf{k}}_2^{(\mathrm{in})}\mathbf{x}\right)}$ of two plane waves with the same neutron energy $\frac{\hbar^2}{2m_{\mathrm{n}}}k_\lambda^2$ in the directions $\hat{\mathbf{k}}_1^{(\mathrm{in})}$ and $\hat{\mathbf{k}}_2^{(\mathrm{in})}$. This initial situation, where the neutron is far away from the scatterer, is represented in (15) by the term $\left\{\mathrm{e}^{i\,k_\lambda\left(\hat{\mathbf{k}}_1^{(\mathrm{in})}\mathbf{x}\right)}+C\mathrm{e}^{i\,k_\lambda\left(\hat{\mathbf{k}}_2^{(\mathrm{in})}\mathbf{x}\right)}\right\}\Psi_\lambda(\mathbf{R})$, which, in accordance with a negligible interaction $V(\mathbf{x},\mathbf{R})$ in the asymptotic regime, is just the (tensor) product of the initial states of the free neutron and of the scattering system. When an incident neutron hits a nucleus in the scatterer, this process becomes the source of scattered outgoing neutron waves of the form $f_{\lambda\lambda'}\left(\mathbf{k}_1^{(\text{in})}, \mathbf{k}_2^{(\text{in})}, \mathbf{k}'; C\right) \frac{\mathrm{e}^{i\,k_{\lambda'}x}}{x}$, where the scattering amplitude $f_{\lambda\lambda'}\left(\mathbf{k}_1^{(\text{in})}, \mathbf{k}_2^{(\text{in})}, \mathbf{k}'; C\right)$, given by (16), depends on the wave number vectors $\mathbf{k}_1^{(\text{in})}$ and $\mathbf{k}_2^{(\text{in})}$ of the incident neutron double-beam, on the initial and final states, $\Psi_\lambda(\mathbf{R})$ and $\Psi_{\lambda'}(\mathbf{R})$, respectively, of the scatterer, on the wave number vector $\mathbf{k}'$ of the scattered neutron, and on C. Also $\frac{\mathrm{e}^{i\,k_{\lambda'}x}}{x}\,\Psi_{\lambda'}(\mathbf{R})$ is a solution of the Schrödinger equation of the total neutron-plus-scatterer system in the absence of an interaction, i.e. $V(\mathbf{x},\mathbf{R}) = 0$, between the neutron and the scatterer, which situation corresponds to the asymptotic regime where the neutron is far away from the scatterer. The scattered state $\frac{\mathrm{e}^{i\,k_{\lambda'}x}}{x}\,\Psi_{\lambda'}(\mathbf{R})$ in the series $\sum_{\lambda'}\cdots$ in (15) is the (tensor) product of an outgoing spherical neutron wave of energy $\frac{\hbar^2}{2m_\text{n}}k_{\lambda'}^2$ and of the final state $\Psi_{\lambda'}(\mathbf{R})$ into which the scatterer transits as a result of the scattering process. Formula (15) for $\Phi_{\text{as}}(\mathbf{x},\mathbf{R})$ contains both forward (for $\hat{\mathbf{k}}' = \mathbf{k}'/|\mathbf{k}'| = \hat{\mathbf{k}}_1^{(\text{in})}$ and $\hat{\mathbf{k}}' = \hat{\mathbf{k}}_2^{(\text{in})}$) and backward (for $\hat{\mathbf{k}}' = -\hat{\mathbf{k}}_1^{(\text{in})}$ and $\hat{\mathbf{k}}' = -\hat{\mathbf{k}}_2^{(\text{in})}$) scattering terms. In the asymptotic regime the asymptotic solution $\Phi_{\text{as}}(\mathbf{x},\mathbf{R})$ of the stationary Schrödinger equation (1) is a linear superposition of the initial and of all possible final scattered states of the whole neutron-plus-scatterer system.

The energy eigenvalue E in (1) is given by

$$E = E_\lambda + \frac{\hbar^2}{2m_\text{n}}k_\text{in}^2 = E_{\lambda'} + \frac{\hbar^2}{2m_\text{n}}k'^2\,, \tag{17}$$

where $k_\text{in} = k_\lambda = \left|\mathbf{k}_1^{(\text{in})}\right| = \left|\mathbf{k}_2^{(\text{in})}\right|$ and $k' = |\mathbf{k}'| = k_{\lambda'}$, rendering the asymptotic form (15) for $\Phi(\mathbf{x},\mathbf{R})$ an eigenfunction of the asymptotic form of the stationary Schrödinger equation (1), i.e. (1) with $V(\mathbf{x},\mathbf{R}) = 0$, with the energy eigenvalue E determined via (17) by the energy eigenvalue E_λ of the initial state $\Psi_\lambda(\mathbf{R})$ of the scatterer (cf. (2)) and by the energy $\frac{\hbar^2}{2m_\text{n}}k_\text{in}^2$ of the incident neutron double-beam. Evidently $\hat{H}_0 - \frac{\hbar^2}{2m_\text{n}}\nabla^2$ is the asymptotic form, in the region where the neutron is far away from the scatterer, of the total neutron-plus-scatterer Hamiltonian on the left of (1). Equation (17) implies that the change in the neutron energy due to scattering is equivalent to the energy transfer from the neutron to the scattering system,

$$\frac{\hbar^2}{2m_\text{n}}\left(k_\text{in}^2 - k'^2\right) = E_{\lambda'} - E_\lambda\,. \tag{18}$$

Assuming now that the initial state is only slightly disturbed by the scattering, one may insert in a first approximation the initial state $\{\mathrm{e}^{i\,k_\text{in}(\hat{\mathbf{k}}_1^{(\text{in})}\mathbf{x}')} + C\mathrm{e}^{i\,k_\text{in}(\hat{\mathbf{k}}_2^{(\text{in})}\mathbf{x}')}\}\Psi_\lambda(\mathbf{R}')$ for $\Phi(\mathbf{x}',\mathbf{R}')$ in the expression (16) for the scattering amplitudes $f_{\lambda\lambda'}\left(\mathbf{k}_1^{(\text{in})}, \mathbf{k}_2^{(\text{in})}, \mathbf{k}'; C\right)$. This approximation is known as the Born approximation [20] and gives rise to the lowest-order term in the Born series for the scattering amplitudes. One then obtains from (16) the result

$$f_{\lambda\lambda'}\left(\mathbf{k}_1^{(\mathrm{in})},\mathbf{k}_2^{(\mathrm{in})},\mathbf{k}';C\right) = -\frac{m_\mathrm{n}}{2\pi\hbar^2}\int\left\{\mathrm{e}^{i\left(\mathbf{k}_1^{(\mathrm{in})}-\mathbf{k}'\right)\mathbf{x}'}+C\mathrm{e}^{i\left(\mathbf{k}_2^{(\mathrm{in})}-\mathbf{k}'\right)\mathbf{x}'}\right\} \times\Psi_{\lambda'}^*(\mathbf{R})V(\mathbf{x}',\mathbf{R})\Psi_\lambda(\mathbf{R})\,d\mathbf{x}'d\mathbf{R}\,. \quad (19)$$

In the Born approximation (19) the scattering amplitude $f_{\lambda\lambda'}\left(\mathbf{k}_1^{(\mathrm{in})},\mathbf{k}_2^{(\mathrm{in})},\mathbf{k}';C\right)$ depends on only the differences $\mathbf{k}_1^{(\mathrm{in})}-\mathbf{k}'$ and $\mathbf{k}_2^{(\mathrm{in})}-\mathbf{k}'$.

To derive an expression for the differential cross section $\frac{d\sigma}{d\Omega}$ of the scattering of a double-beam as described above from a many-body condensed system, we first observe that the probability current density

$$\mathbf{j}(\mathbf{x}) = \frac{i\hbar}{2m_\mathrm{n}}\left[\varphi(\mathbf{x})\nabla\varphi^*(\mathbf{x})-\varphi^*(\mathbf{x})\nabla\varphi(\mathbf{x})\right] \quad (20)$$

for the incident double-beam of neutrons with $\varphi(\mathbf{x})=\mathrm{e}^{i\,k_\lambda\left(\hat{\mathbf{k}}_1^{(\mathrm{in})}\mathbf{x}\right)}+C\mathrm{e}^{i\,k_\lambda\left(\hat{\mathbf{k}}_2^{(\mathrm{in})}\mathbf{x}\right)}$ assumes the form

$$\begin{aligned}\mathbf{j}_\mathrm{in}(\mathbf{x}) &= \frac{\hbar}{m_\mathrm{n}}\left\{\mathbf{k}_1^{(\mathrm{in})}+|C|^2\mathbf{k}_2^{(\mathrm{in})}\right\}\\ &\quad+\frac{\hbar}{2m_\mathrm{n}}\left(\mathbf{k}_1^{(\mathrm{in})}+\mathbf{k}_2^{(\mathrm{in})}\right)\left\{C^*\mathrm{e}^{i\left(\mathbf{k}_1^{(\mathrm{in})}-\mathbf{k}_2^{(\mathrm{in})}\right)\mathbf{x}}+C\mathrm{e}^{-i\left(\mathbf{k}_1^{(\mathrm{in})}-\mathbf{k}_2^{(\mathrm{in})}\right)\mathbf{x}}\right\}\,, \quad (21)\end{aligned}$$

which for $|C|=1$ is parallel to $\mathbf{k}_1^{(\mathrm{in})}+\mathbf{k}_2^{(\mathrm{in})}$. The incident flux $F\left(\mathbf{k}_1^{(\mathrm{in})},\mathbf{k}_2^{(\mathrm{in})};C\right)$ of neutrons crossing per unit time a unit area is then given by $A^{-1}\int_A \mathbf{j}_\mathrm{in}(\mathbf{x})\,d\mathbf{f}$, wherein the surface integral extends over the chosen area A and where the directed surface element $d\mathbf{f}$ is perpendicular to the (planar) surface spanning the area A,

$$F\left(\mathbf{k}_1^{(\mathrm{in})},\mathbf{k}_2^{(\mathrm{in})};C\right) = A^{-1}\int_A \mathbf{j}_\mathrm{in}(\mathbf{x})\,d\mathbf{f}\,. \quad (22)$$

Here we consider in particular a planar surface that is perpendicular to $\mathbf{k}_1^{(\mathrm{in})}+\mathbf{k}_2^{(\mathrm{in})}$. Because $\mathbf{k}_1^{(\mathrm{in})}-\mathbf{k}_2^{(\mathrm{in})}$ is perpendicular to $\mathbf{k}_1^{(\mathrm{in})}+\mathbf{k}_2^{(\mathrm{in})}$, one may represent $\mathbf{k}_1^{(\mathrm{in})}-\mathbf{k}_2^{(\mathrm{in})}=\tilde{\mathbf{k}}$ in the form $\tilde{\mathbf{k}}=\left(\tilde{k}_{x'},\tilde{k}_{y'},0\right)$, where $\tilde{k}_{x'}$ and $\tilde{k}_{y'}$ are the x'- and y'-components, respectively, of $\tilde{\mathbf{k}}$ in a right-handed Cartesian $x'y'z'$-frame of reference whose z'-axis is pointing in the direction of $\mathbf{k}_1^{(\mathrm{in})}+\mathbf{k}_2^{(\mathrm{in})}$. Considering then an area spanning in the x'- and y'-directions a length $2\pi/\tilde{k}_{x'}$ and $2\pi/\tilde{k}_{y'}$, respectively, one then immediately obtains from (22) the result

$$F\left(\mathbf{k}_1^{(\mathrm{in})},\mathbf{k}_2^{(\mathrm{in})};C\right) = \frac{\hbar}{m_\mathrm{n}}\sqrt{\tfrac{1}{2}\left(k_\mathrm{in}^2+\mathbf{k}_1^{(\mathrm{in})}\mathbf{k}_2^{(\mathrm{in})}\right)}\left(1+|C|^2\right) \quad (23)$$

for the incident flux $F\left(\mathbf{k}_1^{(\mathrm{in})},\mathbf{k}_2^{(\mathrm{in})};C\right)$ of neutrons per unit area and per unit time through such a surface perpendicular to $\mathbf{k}_1^{(\mathrm{in})}+\mathbf{k}_2^{(\mathrm{in})}$. For $\mathbf{k}_2^{(\mathrm{in})}=\mathbf{k}_1^{(\mathrm{in})}=\mathbf{k}_\mathrm{in}$

and $C = 0$ one retrieves from (23) the familiar result $(\hbar/m_{\rm n})k_{\rm in}$ ($k_{\rm in} = |\mathbf{k}_{\rm in}|$) for the incoming neutron flux of a conventional single-beam [20].

From the above general formula (20) for the probability current density $\mathbf{j}(\mathbf{x})$ one also obtains

$$\mathbf{j}_{\rm out}(\mathbf{x}) = \frac{\hbar}{m_{\rm n}}\frac{k'}{x^2}\left|f_{\lambda\lambda'}\left(\mathbf{k}_1^{(\rm in)}, \mathbf{k}_2^{(\rm in)}, \mathbf{k}'; C\right)\right|^2 \hat{\mathbf{x}} \tag{24}$$

for the probability current densities of the outgoing neutron waves in (15). The flux of neutrons scattered per unit time into the solid angle element $d\Omega$ is then given by

$$\mathbf{j}_{\rm out}(\mathbf{x})x^2\hat{\mathbf{x}}\, d\Omega = \frac{\hbar}{m_{\rm n}}k'\left|f_{\lambda\lambda'}\left(\mathbf{k}_1^{(\rm in)}, \mathbf{k}_2^{(\rm in)}, \mathbf{k}'; C\right)\right|^2 d\Omega\,. \tag{25}$$

With these findings the differential cross section may be written in the form

$$\begin{aligned}\frac{d\sigma}{d\Omega} &= \frac{\frac{\hbar}{m_{\rm n}}k'}{F\left(\mathbf{k}_1^{(\rm in)}, \mathbf{k}_2^{(\rm in)}; C\right)}\sum_{\lambda\lambda'}\frac{\mathrm{e}^{-\beta E_\lambda}}{Z(\beta)}\left|f_{\lambda\lambda'}\left(\mathbf{k}_1^{(\rm in)}, \mathbf{k}_2^{(\rm in)}, \mathbf{k}'; C\right)\right|^2 \\ &= Nb^2\frac{\frac{\hbar}{m_{\rm n}}k'}{F\left(\mathbf{k}_1^{(\rm in)}, \mathbf{k}_2^{(\rm in)}; C\right)}\mathcal{D}(\mathbf{k}_1, \mathbf{k}_2; C)\,,\end{aligned} \tag{26}$$

where

$$\mathcal{D}(\mathbf{k}_1, \mathbf{k}_2; C) = \frac{1}{Nb^2}\sum_{\lambda\lambda'}\frac{\mathrm{e}^{-\beta E_\lambda}}{Z(\beta)}\left|f_{\lambda\lambda'}\left(\mathbf{k}_1^{(\rm in)}, \mathbf{k}_2^{(\rm in)}, \mathbf{k}'; C\right)\right|^2 . \tag{27}$$

In (26) and (27) the quantity

$$Z(\beta) = \sum_\lambda \mathrm{e}^{-\beta E_\lambda} \tag{28}$$

is the canonical partition function of the scattering many-body system of N particles at temperature T and $\beta = 1/k_{\rm B}T$, with the Boltzmann constant $k_{\rm B}$, is the unitless inverse temperature. The letter b in (26) and (27) denotes the neutron scattering length, to be explained further below. The neutron linear momenta transfers to the scatterer are given by

$$\hbar\mathbf{k}_1 = \hbar\left(\mathbf{k}_1^{(\rm in)} - \mathbf{k}'\right) \quad \text{and} \quad \hbar\mathbf{k}_2 = \hbar\left(\mathbf{k}_2^{(\rm in)} - \mathbf{k}'\right)\,. \tag{29}$$

The double series $\sum_{\lambda\lambda'}\cdots$ in (26) and (27) extends over all final states $\Psi_{\lambda'}(\mathbf{R})$ into which the scatterer can transit as a result of the scattering and over all initial states $\Psi_\lambda(\mathbf{R})$ from which scattering can occur. However, at temperature $T = 1/\beta k_{\rm B}$ the scatterer's initial states $\Psi_\lambda(\mathbf{R})$ are occupied with probabilities $\mathrm{e}^{-\beta E_\lambda}/Z(\beta)$, which explains the series over all λ in addition to the ones over all λ' and the occupation probability factors in (26) and (27).

Equation (26) for the differential cross section $\frac{d\sigma}{d\Omega}$ is valid in the static approximation [20], in which it is assumed that the scattering is nearly elastic. For an incident neutron energy very much in excess of the energy eigenvalues

$E_{\lambda'}$ of the target many-body system the change in the neutron's energy is very small and the scattering appears to be elastic to a good approximation. However, truly elastic scattering is distinct from nearly elastic scattering, leading to the static approximation result (26) for the differential cross section $\frac{d\sigma}{d\Omega}$ of scattering of an incident double-beam of neutrons from a condensed quantum many-body system.

3 The Bragg and Diffuse Scattering Structure Functions

In this section we will express the total scattering structure factor $\mathcal{D}(\mathbf{k}_1, \mathbf{k}_2; C)$, defined in (27), in terms of the local one- and two-body densities of the scattering material.

Inserting the result (19) for the scattering amplitudes $f_{\lambda\lambda'}\left(\mathbf{k}_1^{(\mathrm{in})}, \mathbf{k}_2^{(\mathrm{in})}, \mathbf{k}'; C\right)$ in the expression (27) for $\mathcal{D}(\mathbf{k}_1, \mathbf{k}_2; C)$ one finds after some rather straightforward algebra

$$\begin{aligned}\mathcal{D}(\mathbf{k}_1, \mathbf{k}_2; C) &= \frac{1}{Nb^2}\left(\frac{m_\mathrm{n}}{2\pi\hbar^2}\right)^2 \int \left\{\mathrm{e}^{i\,\mathbf{k}_1\mathbf{x}_1} + C\mathrm{e}^{i\,\mathbf{k}_2\mathbf{x}_2}\right\}\left\{\mathrm{e}^{-i\,\mathbf{k}_1\mathbf{x}_1} + C^*\mathrm{e}^{-i\,\mathbf{k}_2\mathbf{x}_2}\right\} \\ &\qquad \times W(\mathbf{R}, \mathbf{R}; \beta) V(\mathbf{x}_1, \mathbf{R}) V(\mathbf{x}_2, \mathbf{R})\, d\mathbf{R} d\mathbf{x}_1 d\mathbf{x}_2 \,.\end{aligned} \tag{30}$$

In deriving (30) we have made use of the fact that the coordinate space matrix elements $W(\mathbf{R}, \mathbf{R}'; \beta)$ of the canonical density operator of the scattering many-body system are given by

$$W(\mathbf{R}, \mathbf{R}'; \beta) = \sum_\lambda \frac{\mathrm{e}^{-\beta E_\lambda}}{Z(\beta)} \Psi_\lambda(\mathbf{R}) \Psi_{\lambda'}^*(\mathbf{R}') \tag{31}$$

and of the completeness relation

$$\sum_{\lambda'} \Psi_{\lambda'}^*(\mathbf{R}) \Psi_{\lambda'}(\mathbf{R}') = \delta(\mathbf{R} - \mathbf{R}') \tag{32}$$

of the eigenstates $\Psi_\lambda(\mathbf{R})$ of the scatterer (cf. (2)), wherein $\delta(\mathbf{R} - \mathbf{R}')$ is the $3N$-dimensional Dirac δ-distribution. Accordingly, only the diagonal coordinate space density matrix elements $W(\mathbf{R}, \mathbf{R}; \beta)$ of the density operator of the scatterer appear in (30). In coordinate space representation the unit-normalization

$$\sum_\lambda \mathrm{e}^{-\beta E_\lambda} / Z(\beta) = 1 \tag{33}$$

of the occupation probabilities $\mathrm{e}^{-\beta E_\lambda} / Z(\beta)$ assumes the form

$$\int W(\mathbf{R}, \mathbf{R}; \beta)\, d\mathbf{R} = 1 \,. \tag{34}$$

The interaction $V(\mathbf{x},\mathbf{R})$ between the neutron and the scatterer may be described by means of the potential

$$V(\mathbf{x},\mathbf{R}) = \frac{2\pi\hbar^2}{m_\mathrm{n}} b \sum_{i=1}^{N} \delta(\mathbf{x}-\mathbf{r}_i)\,, \tag{35}$$

where the sum extends over all N particles in the condensed many-body system. In this work we consider only scattering systems made out of identical particles. The neutron scattering length b, which is characteristic for the scattering material, thus turns out to be basically the strength of the δ-interaction in (35) by means of which the neutron at $\mathbf{x}$ and particle i of the scatterer at $\mathbf{r}_i$ interact with each other. Such a δ-interaction model of the neutron-scatterer potential $V(\mathbf{x},\mathbf{R})$ is reasonable because of the extreme short-range nature of the nucleonic force via which the neutron interacts with the scatterer's particles' nuclei. For the same reason $V(\mathbf{x},\mathbf{R})$ may be written as a sum of individual neutron-particle interactions. Neutrons are scattered by the nuclei in the scattering condensed matter system. When adopting a neutron-scatterer interaction $V(\mathbf{x},\mathbf{R})$ of the form (35), which is the sum of the potentials between the incident neutron and the individual nuclei in the target system, it is further assumed that the incoming neutron collides with just one atom in the scatterer and is not scattered a second time by another particle in the target many-body system before escaping to a large distance from the target [20]. This assumption is valid, if the scattering amplitude is small, which condition justifies the use of the Born approximation [20]. Here we do not consider processes such as double scattering.

Using the neutron-scatterer potential (35) in (30) one derives with some fairly straightforward algebraic manipulations the formula

$$\begin{aligned}\mathcal{D}(\mathbf{k}_1,\mathbf{k}_2;C) = 1 &+|C|^2 + \varrho_0^{-1}\left\{C\rho(\mathbf{k}_1-\mathbf{k}_2) + C^*\rho(\mathbf{k}_2-\mathbf{k}_1)\right\}\\ &+\frac{1}{N}\int \left\{\mathrm{e}^{i\,\mathbf{k}_1\mathbf{x}_1} + C\mathrm{e}^{i\,\mathbf{k}_2\mathbf{x}_1}\right\}\left\{\mathrm{e}^{-i\,\mathbf{k}_1\mathbf{x}_2} + C^*\mathrm{e}^{-i\,\mathbf{k}_2\mathbf{x}_2}\right\}\\ &\quad\times\rho_2(\mathbf{x}_1,\mathbf{x}_2)\,d\mathbf{x}_1 d\mathbf{x}_2\,,\end{aligned} \tag{36}$$

where the two-body density $\rho_2(\mathbf{x}_1,\mathbf{x}_2)$ is defined as

$$\begin{aligned}\rho_2(\mathbf{x}_1,\mathbf{x}_2) &= \sum_{i\neq j=1}^{N}\int W(\mathbf{R},\mathbf{R};\beta)\delta(\mathbf{x}_1-\mathbf{r}_i)\delta(\mathbf{x}_2-\mathbf{r}_j)\,d\mathbf{R}\\ &= N(N-1)\int W(\mathbf{x}_1,\mathbf{x}_2,\mathbf{r}_3,\cdots,\mathbf{r}_N,\mathbf{x}_1,\mathbf{x}_2,\mathbf{r}_3,\cdots,\mathbf{r}_N;\beta)\,d\mathbf{r}_3\cdots d\mathbf{r}_N\,.\end{aligned} \tag{37}$$

The double sum $\sum_{i\neq j=1}^{N}\cdots$ in the definition (37) extends over all $N(N-1)$ pairs of different particles i and j, i.e. $i=j$ terms are excluded from the double sum in (37). The quantity $\varrho_0 = N/V$ in (36) is simply the bulk particle number density of the many-body system of N particles, where V is the volume occupied by the scatterer. In (36)

$$\rho(\mathbf{k}'') = \frac{1}{V}\int_V \varrho(\mathbf{x})\mathrm{e}^{-i\,\mathbf{k}''\mathbf{x}}\,d\mathbf{x} = \frac{1}{\Omega}\int_\Omega \varrho(\mathbf{x})\mathrm{e}^{-i\,\mathbf{k}''\mathbf{x}}\,d\mathbf{x} \tag{38}$$

with $\mathbf{k}'' = \pm(\mathbf{k}_1 - \mathbf{k}_2)$ is the Fourier transform of the one-body density $\varrho(\mathbf{x})$, which in turn is defined as

$$\begin{aligned}\varrho(\mathbf{x}) &= \sum_{i=1}^{N} \int W(\mathbf{R},\mathbf{R};\beta)\delta(\mathbf{x}-\mathbf{r}_i)\, d\mathbf{R} \\ &= N \int W(\mathbf{x},\mathbf{r}_2,\cdots,\mathbf{r}_N,\mathbf{x},\mathbf{r}_2\cdots,\mathbf{r}_N;\beta)\, d\mathbf{r}_2\cdots d\mathbf{r}_N\,. \end{aligned} \tag{39}$$

The integral $\int_V \cdots$ in (38) extends over the volume V occupied by the scattering system. In the case of crystalline systems $\rho(\mathbf{k}'')$ may be written in the second form, a lattice Fourier transform, given in (38), where the integral $\int_\Omega \cdots$ is over the volume Ω of a primitive unit cell of the Bravais lattice of the crystal. We assume that the crystalline scatterer occupies a volume V that consists of an integral number of primitive unit cells of volume Ω. To simplify the notation, we do not explicitly indicate by an argument T or β the temperature-dependence of $\varrho(\mathbf{x})$, $\rho(\mathbf{k}'')$, and $\rho_2(\mathbf{x}_1,\mathbf{x}_2)$ or of other quantities introduced above and further below. However, all these quantities do depend on temperature. In passing we note that the one-body density $\varrho(\mathbf{x})$ is the number of particles per unit volume at $\mathbf{x} \in |\mathrm{R}^3$, whereas the two-body density $\rho_2(\mathbf{x}_1,\mathbf{x}_2)$ describes the probability for a particle to be at a position $\mathbf{x}_1 \in |\mathrm{R}^3$ in three-dimensional space $|\mathrm{R}^3$ and another one simultaneously at a position $\mathbf{x}_2 \in |\mathrm{R}^3$.

To make contact with previous works, we exploit the Fourier series representation

$$\varrho_2(\mathbf{S},\mathbf{r}) = \rho_2\left(\mathbf{S}+\tfrac{1}{2}\mathbf{r},\mathbf{S}-\tfrac{1}{2}\mathbf{r}\right) = \sum_{\mathbf{K}'} u(\mathbf{r};\mathbf{K}')\mathrm{e}^{i\mathbf{K}'\mathbf{S}} \tag{40}$$

of the two-body density $\varrho_2(\mathbf{S},\mathbf{r})$ given as defined above as a function of the center-of-mass position $\mathbf{S}$ and of the relative position vector $\mathbf{r}$,

$$\mathbf{S} = \tfrac{1}{2}(\mathbf{x}_1+\mathbf{x}_2) \quad \text{and} \quad \mathbf{r} = \mathbf{x}_1 - \mathbf{x}_2\,, \tag{41}$$

of the two particles at $\mathbf{x}_1$ and $\mathbf{x}_2$. Because here we consider a crystalline scatterer, the series in (40) extends over all lattice vectors $\mathbf{K}'$ of the reciprocal lattice of the direct Bravais lattice of the crystal. The lattice Fourier transform $u(\mathbf{r};\mathbf{K}')$ of the two-body density $\varrho_2(\mathbf{S},\mathbf{r})$ in (40) is given by

$$u(\mathbf{r};\mathbf{K}') = \frac{1}{\Omega}\int_\Omega \varrho_2(\mathbf{S},\mathbf{r})\mathrm{e}^{-i\mathbf{K}'\mathbf{S}}\, d\mathbf{S}\,, \tag{42}$$

wherein the integration extends over the volume Ω of a primitive unit cell.

With the help of (40)–(42) formula (36) may be cast in the shape

$$\begin{aligned}\mathcal{D}(\mathbf{k}_1,\mathbf{k}_2;C) = 1 &+ |C|^2 + \varrho_0^{-1}\left\{C\rho(\mathbf{K}) + C^*\rho^*(\mathbf{K})\right\} \\ &+ \varrho_0^{-1}\mathcal{F}\left[u(\mathbf{r};0)\right](\mathbf{k}_1) + |C|^2\mathcal{F}\left[u(\mathbf{r};0)\right](\mathbf{k}_2) \\ &+ \varrho_0^{-1}\left\{C\mathcal{F}\left[u(\mathbf{r};\mathbf{K})\right](\mathbf{k}) + C^*\mathcal{F}^*\left[u(\mathbf{r};\mathbf{K})\right](\mathbf{k})\right\}\,, \end{aligned} \tag{43}$$

where

$$\mathbf{K} = \mathbf{k}_1 - \mathbf{k}_2 = \mathbf{k}_1^{(\mathrm{in})} - \mathbf{k}_2^{(\mathrm{in})} \tag{44}$$

and

$$\mathbf{k} = \tfrac{1}{2}(\mathbf{k}_1 + \mathbf{k}_2) = \tfrac{1}{2}\left(\mathbf{k}_1^{(\mathrm{in})} + \mathbf{k}_2^{(\mathrm{in})}\right) - \mathbf{k}' . \tag{45}$$

The continuous Fourier transforms $\mathcal{F}[u(\mathbf{r};0)](\mathbf{k}_1)$, $\mathcal{F}[u(\mathbf{r};0)](\mathbf{k}_2)$, and $\mathcal{F}[u(\mathbf{r};\mathbf{K})](\mathbf{k})$ of the discrete lattice Fourier transform functions $u(\mathbf{r};0)$ and $u(\mathbf{r};\mathbf{K})$ (cf. (40) and (42)) in (43) may be obtained as

$$\mathcal{F}[u(\mathbf{r};\mathbf{K})](\mathbf{k}) = \int u(\mathbf{r};\mathbf{K}) \mathrm{e}^{-i\,\mathbf{kr}}\, d\mathbf{r} . \tag{46}$$

In deriving the result (43) we have also made use of the time-reversal symmetry $\rho(-\mathbf{K}) = \rho^*(\mathbf{K})$ in the one-body density and of the exchange symmetry $\rho_2(\mathbf{x}_1,\mathbf{x}_2) = \rho_2(\mathbf{x}_2,\mathbf{x}_1)$ or, equivalently, $u(-\mathbf{r};\mathbf{K}) = u(\mathbf{r};\mathbf{K})$ for all $\mathbf{K}$ in the two-body density. It is important to realize that for the $\mathbf{K}$-dependent terms in (43) to yield non-vanishing contributions to $\mathcal{D}(\mathbf{k}_1,\mathbf{k}_2;C)$ the difference $\mathbf{K} = \mathbf{k}_1 - \mathbf{k}_2$ of the momenta transfers $\mathbf{k}_1$ and $\mathbf{k}_2$ must be a reciprocal lattice vector of the scattering crystal. Equation (44) implies that the two wave number vectors $\mathbf{k}_1^{(\mathrm{in})}$ and $\mathbf{k}_2^{(\mathrm{in})}$ in the incident neutron double-beam must also differ by a reciprocal lattice vector $\mathbf{K}$ of the scattering crystal. However, the vector $\mathbf{k}$ in (45) is not subject to any such requirements for $\mathcal{F}[u(\mathbf{r};\mathbf{K})](\mathbf{k})$ to be non-zero.

Further valuable insight into the physical nature of the scattering quantity $\mathcal{D}(\mathbf{k}_1,\mathbf{k}_2;C)$ may be gained by splitting it further up into leading-order Bragg scattering and diffuse background scattering terms. Starting from (43) the final result of this analysis reads

$$\begin{aligned}\mathcal{D}(\mathbf{k}_1,\mathbf{k}_2;C) = {} & V\{S_\mathrm{B}(0;\mathbf{k}_1) + |C|^2 S_\mathrm{B}(0;\mathbf{k}_2) + C S_\mathrm{B}(\mathbf{K};\mathbf{k}) + C^* S_\mathrm{B}^*(\mathbf{K};\mathbf{k})\} \\ & + S_\mathrm{d}(0;\mathbf{k}_1) + |C|^2 S_\mathrm{d}(0;\mathbf{k}_2) + C S_\mathrm{d}(\mathbf{K};\mathbf{k}) + C^* S_\mathrm{d}^*(\mathbf{K};\mathbf{k}) \end{aligned} \tag{47}$$

with $\mathbf{K}$ and $\mathbf{k}$ as in (44) and (45). The Bragg scattering structure function $S_\mathrm{B}(\mathbf{K};\mathbf{k})$ in this equation is given by

$$S_\mathrm{B}(\mathbf{K};\mathbf{k}) = \varrho_0^{-1} \rho\left(\mathbf{k} + \tfrac{1}{2}\mathbf{K}\right) \rho^*\left(\mathbf{k} - \tfrac{1}{2}\mathbf{K}\right) \tag{48}$$

and because of the prefactor V, by which all Bragg scattering structure functions in the expression (47) for $\mathcal{D}(\mathbf{k}_1,\mathbf{k}_2;C)$ are multiplied, yields scattering terms which are on the order of the volume V occupied by the scattering sample. The diffuse scattering structure function $S_\mathrm{d}(\mathbf{K};\mathbf{k})$ in (47) assumes the shape

$$S_\mathrm{d}(\mathbf{K};\mathbf{k}) = \varrho_0^{-1}\{\rho(\mathbf{K}) + \mathcal{F}[u(\mathbf{r};\mathbf{K})](\mathbf{k})\} - V S_\mathrm{B}(\mathbf{K};\mathbf{k}) . \tag{49}$$

Note that $\rho(0) = \varrho_0$ and

$$\mathbf{k}_1 = \mathbf{k} + \tfrac{1}{2}\mathbf{K} \quad \text{and} \quad \mathbf{k}_2 = \mathbf{k} - \tfrac{1}{2}\mathbf{K} . \tag{50}$$

Observing that the diffuse scattering structure function
$S_\mathrm{d}(\mathbf{K};\mathbf{k})$ defined in (49) may be written also as

$$\begin{aligned} S_\mathrm{d}(\mathbf{K};\mathbf{k}) &= \frac{\rho(\mathbf{K})}{\varrho_0} + \frac{1}{N}\int \left[\varrho_2(\mathbf{S},\mathbf{r}) - \varrho\left(\mathbf{S} + \tfrac{1}{2}\mathbf{r}\right)\varrho\left(\mathbf{S} - \tfrac{1}{2}\mathbf{r}\right)\right] \mathrm{e}^{-i\,\mathbf{kr}} \mathrm{e}^{-i\,\mathbf{KS}}\, d\mathbf{r} d\mathbf{S} \\ &= \frac{\rho(\mathbf{K})}{\varrho_0} + \frac{1}{N}\int \left[\rho_2(\mathbf{x}_1,\mathbf{x}_2) - \varrho(\mathbf{x}_1)\varrho(\mathbf{x}_2)\right] \mathrm{e}^{-i\,(\mathbf{k}_1\mathbf{x}_1 - \mathbf{k}_2\mathbf{x}_2)}\, d\mathbf{x}_1 d\mathbf{x}_2 , \end{aligned} \tag{51}$$

one finds that the diffuse scattering function $S_{\mathrm{d}}(\mathbf{K};\mathbf{k})$ is on the order $\mathcal{O}(1)$. Because of the asymptotic behavior

$$\lim_{|\mathbf{r}|\longrightarrow\infty} \varrho_2(\mathbf{S},\mathbf{r}) = \varrho\left(\mathbf{S}+\tfrac{1}{2}\mathbf{r}\right)\varrho\left(\mathbf{S}-\tfrac{1}{2}\mathbf{r}\right) = \varrho(\mathbf{x}_1)\varrho(\mathbf{x}_2) = \lim_{|\mathbf{x}_1-\mathbf{x}_2|\longrightarrow\infty} \rho_2(\mathbf{x}_1,\mathbf{x}_2) \tag{52}$$

of the two-body density the double integrals in (51) are on the order V of the volume of the scatterer. Due to the division by the total number N of particles in the scatterer the contributions $N^{-1}\int\cdots = \varrho_0^{-1}V^{-1}\int\cdots$ in (51) are on the order $\mathcal{O}(1)$, rendering $S_{\mathrm{d}}(\mathbf{K};\mathbf{k})$ on the order $\mathcal{O}(1)$. Whence the contributions to the total scattering function $\mathcal{D}(\mathbf{k}_1,\mathbf{k}_2;C)$ in (47) stemming from the diffuse scattering function (49) are on the order $\mathcal{O}(1)$ in contrast to the dominant leading-order Bragg scattering peaks, which are volume-extensive. Equation (52) shows that the joint probability $\rho_2(\mathbf{x}_1,\mathbf{x}_2)$, the two-body density, for a particle to be at $\mathbf{x}_1$ and another one simultaneously at $\mathbf{x}_2$ becomes the product $\varrho(\mathbf{x}_1)\varrho(\mathbf{x}_2)$ of the uncorrelated probabilities $\varrho(\mathbf{x}_1)$ and $\varrho(\mathbf{x}_2)$ in the limit of large interparticle distances $|\mathbf{r}| = |\mathbf{x}_1-\mathbf{x}_2| \longrightarrow \infty$. Equations (36), (43), and in particular (47), (49), (51), and (52) show that the dominant volume-extensive Bragg scattering peaks in the total (Bragg plus diffuse) scattering structure function $\mathcal{D}(\mathbf{k}_1,\mathbf{k}_2;C)$, (47), have their origin actually in the asymptotic long-range behavior (52) of the two-body distribution function $\varrho_2(\mathbf{S},\mathbf{r})$. Because, as will be shown in the next section, for $\mathbf{K}=0$ the Bragg and the diffuse scattering structure functions $S_{\mathrm{B}}(0;\mathbf{k})$ and $S_{\mathrm{d}}(0;\mathbf{k})$ are real, also the total scattering structure function $\mathcal{D}(\mathbf{k}_1,\mathbf{k}_2;C)$ in (47) and the differential cross section $\frac{d\sigma}{d\Omega}$ in (26) are real quantities.

The terms in (47) that depend on only $\mathbf{k}_1$ or $\mathbf{k}_2$ evidently are the Bragg and diffuse scattering terms from the scattering of the single beams $\mathrm{e}^{i\,\mathbf{k}_1^{(\mathrm{in})}\mathbf{x}}$ and $\mathrm{e}^{i\,\mathbf{k}_2^{(\mathrm{in})}\mathbf{x}}$ in the incoming double-beam $\mathrm{e}^{i\,\mathbf{k}_1^{(\mathrm{in})}\mathbf{x}} + C\mathrm{e}^{i\,\mathbf{k}_2^{(\mathrm{in})}\mathbf{x}}$. Such contributions in the total differential cross section (26) yield access only to the magnitudes $|\rho(\mathbf{k}_1)|$ and $|\rho(\mathbf{k}_2)|$ of the Fourier transform of the one-body density $\varrho(\mathbf{x})$ from the Bragg scattering peaks $VS_{\mathrm{B}}(0;\mathbf{k}_1)$ and $VS_{\mathrm{B}}(0;\mathbf{k}_2)$ and, from the diffuse background scattering, to the Fourier transform $\mathcal{F}\left[u(\mathbf{r};0)\right](\mathbf{k}_1)$ and $\mathcal{F}\left[u(\mathbf{r};0)\right](\mathbf{k}_2)$ of the lowest-lying $\mathbf{K}=0$ term $u(\mathbf{r};0)$ in the lattice Fourier series expansion (40) of the full two-body density $\varrho_2(\mathbf{S},\mathbf{r})$. However, to obtain this information about the microstructure of the scattering material a two-beam experiment is not needed, since the same information may be extracted already from the scattering cross section of a conventional single-beam experiment. It is trivial to prove that (47) yields the total (Bragg plus diffuse) scattering structure function for the scattering of an incident single beam of the form $\mathrm{e}^{i\,\mathbf{k}_1^{(\mathrm{in})}\mathbf{x}}$, i.e. $C=0$. It then is clear that the terms in (47) that depend on only $\mathbf{k}_1$ or $\mathbf{k}_2$ do not yield any further information that a conventional single-beam experiment would not give already. It is also obvious that the Bragg scattering structure functions $S_{\mathrm{B}}(0;\mathbf{k}_1)$ and $S_{\mathrm{B}}(0;\mathbf{k}_2)$ in (47) can be non-zero only for $\mathbf{k}_1$ and $\mathbf{k}_2$ reciprocal lattice vectors of the scattering crystal.

To relate also the higher-lying $\mathbf{K}\neq 0$ terms $u(\mathbf{r};\mathbf{K})$ in the expansion (40) of the full two-body density $\varrho_2(\mathbf{S},\mathbf{r})$ to experiment, a proper two-beam scattering experiment with an incident double-beam of the form $\mathrm{e}^{i\,\mathbf{k}_1^{(\mathrm{in})}\mathbf{x}} + C\mathrm{e}^{i\,\mathbf{k}_2^{(\mathrm{in})}\mathbf{x}}$ with

$C \neq 0$ is needed. The terms in (47) that depend on both $\mathbf{k}_1$ and $\mathbf{k}_2$ (cf. (44), (45), and (50)) are the genuine double-beam scattering contributions, stemming from the interference of the scattering of $\mathrm{e}^{i\,\mathbf{k}_1^{(\mathrm{in})}\mathbf{x}}$ and $\mathrm{e}^{i\,\mathbf{k}_2^{(\mathrm{in})}\mathbf{x}}$ in the incoming superposition $\mathrm{e}^{i\,\mathbf{k}_1^{(\mathrm{in})}\mathbf{x}} + C\mathrm{e}^{i\,\mathbf{k}_2^{(\mathrm{in})}\mathbf{x}}$. Also these terms consist of leading-order Bragg peaks, proportional to the size of the scattering sample and described by the Bragg scattering structure function (48), and of diffuse background scattering contributions, characterized by the diffuse scattering structure function (49). The proper two-beam Bragg scattering peaks can be non-zero only for both momenta transfers $\mathbf{k} + \frac{1}{2}\mathbf{K} = \mathbf{k}_1$ and $\mathbf{k} - \frac{1}{2}\mathbf{K} = \mathbf{k}_2$ to the scatterer being reciprocal lattice vectors of the scattering crystal. Else there is only the diffuse background. However, we emphasize that for the diffuse genuine double-beam contribution in (47), given by the sum of diffuse scattering terms $CS_\mathrm{d}(\mathbf{K};\mathbf{k}) + C^* S_\mathrm{d}^*(\mathbf{K};\mathbf{k})$, depending on both $\mathbf{K}$ and $\mathbf{k}$, to be non-zero at all, the difference $\mathbf{k}_1 - \mathbf{k}_2$ of the momenta transfers and whence via (44) the difference $\mathbf{k}_1^{(\mathrm{in})} - \mathbf{k}_2^{(\mathrm{in})}$ of the wave number vectors in the incident two-beam $\mathrm{e}^{i\,\mathbf{k}_1^{(\mathrm{in})}\mathbf{x}} + C\mathrm{e}^{i\,\mathbf{k}_2^{(\mathrm{in})}\mathbf{x}}$ must be a reciprocal lattice vector $\mathbf{K}$ of the scattering crystal. Obviously, this $\mathbf{K}$ must be different from zero, since $\mathbf{K} = 0$ corresponds to a conventional single-beam experiment with an incoming beam $(1 + C)\mathrm{e}^{i\,\mathbf{k}_1^{(\mathrm{in})}\mathbf{x}}$.

To obtain information about $u(\mathbf{r};\mathbf{K})$ for $\mathbf{K} \neq 0$ in the form $CS_\mathrm{d}(\mathbf{K};\mathbf{k}) + C^* S_\mathrm{d}(\mathbf{K};\mathbf{k})$ in (47) from a double-beam experiment, one would have to first separate from the total (Bragg plus diffuse) scattering cross section the diffuse background scattering terms, from which one may then obtain the total diffuse scattering term $S_\mathrm{d}(0;\mathbf{k}_1) + |C|^2 S_\mathrm{d}(0;\mathbf{k}_2) + CS_\mathrm{d}(\mathbf{K};\mathbf{k}) + C^* S_\mathrm{d}(\mathbf{K};\mathbf{k})$ in (47). Since $S_\mathrm{d}(0;\mathbf{k}_1)$ and $S_\mathrm{d}(0;\mathbf{k}_2)$ would be known from the diffuse background scattering in conventional single-beam experiments, one then has $CS_\mathrm{d}(\mathbf{K};\mathbf{k}) + C^* S_\mathrm{d}(\mathbf{K};\mathbf{k})$ as mentioned above.

It is interesting to see what (47)–(49) yield in the case of a spatially uniform system, such as a liquid of bulk density ϱ_0. In such a case we have $\rho(\mathbf{K}) = 0$ and $u(\mathbf{r};\mathbf{K}) = 0$ for all $\mathbf{K} \neq 0$ and

$$\begin{aligned}\mathcal{D}_{\mathrm{uniform}}(\mathbf{k}_1, \mathbf{k}_2; C) = N\{&\delta_{\mathbf{k}_1 0} + |C|^2 \delta_{\mathbf{k}_2 0} + 2\mathrm{Re}\{C\}\delta_{\mathbf{k}_1 0}\delta_{\mathbf{k}_2 0}\} \\ &+ S(k_1) + |C|^2 S(k_2) + 2\delta_{\mathbf{K}0}\mathrm{Re}\{C\} S(k)\,, \end{aligned} \tag{53}$$

where $k_1 = |\mathbf{k}_1|$, $k_2 = |\mathbf{k}_2|$, $k = |\mathbf{k}|$, and where

$$S(k) = 1 + \varrho_0^{-1} \int [g(r) - 1]\mathrm{e}^{-i\,\mathbf{k}\mathbf{r}}\, d\mathbf{r} = 1 + \frac{4\pi}{k}\varrho_0^{-1} \int_0^\infty r[g(r) - 1]\sin(kr)\, dr \tag{54}$$

is the usual liquid structure function. For a spatially uniform system one has $\varrho_2(\mathbf{S},\mathbf{r}) = u(\mathbf{r};0) = \varrho_0^2 g(r)$ $(r = |\mathbf{r}|)$, where $g(r)$ is the radial distribution function. The only Bragg scattering peaks in (53) appear in the forward directions for $\mathbf{k}' = \mathbf{k}_1^{(\mathrm{in})}$ and/or $\mathbf{k}' = \mathbf{k}_2^{(\mathrm{in})}$. In all other directions there are diffuse scattering terms only. In this sense there is no Bragg scattering but only diffuse scattering from a uniform liquid. Evidently, for a spatially uniform system a two-beam experiment cannot lead to any further information about the structure than can be obtained already from conventional single-beam scattering.

The result (47) may be also written in the form

$$\mathcal{D}(\mathbf{k}_1,\mathbf{k}_2;C) = \frac{1}{N}\left\langle\left|\sum_{i=1}^{N}\left\{e^{i\,\mathbf{k}_1\mathbf{r}_i} + Ce^{i\,\mathbf{k}_2\mathbf{r}_i}\right\}\right|^2\right\rangle$$
$$= \frac{1}{N}\left|\left\langle\sum_{i=1}^{N}\left\{e^{i\,\mathbf{k}_1\mathbf{r}_i} + Ce^{i\,\mathbf{k}_2\mathbf{r}_i}\right\}\right\rangle\right|^2$$
$$+\frac{1}{N}\left\{\left\langle\left|\sum_{i=1}^{N}\left\{e^{i\,\mathbf{k}_1\mathbf{r}_i} + Ce^{i\,\mathbf{k}_2\mathbf{r}_i}\right\}\right|^2\right\rangle - \left|\left\langle\sum_{i=1}^{N}\left\{e^{i\,\mathbf{k}_1\mathbf{r}_i} + Ce^{i\,\mathbf{k}_2\mathbf{r}_i}\right\}\right\rangle\right|^2\right\}, \tag{55}$$

where $\langle\cdots\rangle = \int \cdots W(\mathbf{R},\mathbf{R};\beta)\,d\mathbf{R}$ with $W(\mathbf{R},\mathbf{R};\beta)$ as in (31) (for $\mathbf{R}' = \mathbf{R} = (\mathbf{r}_1, \mathbf{r}_2, \cdots, \mathbf{r}_N)$ therein) denotes the thermodynamic expectation value w.r.t. the canonical density operator of the scatterer. It is fairly straightforward to show that in the expression (55) the square of the absolute value of the thermodynamic expectation value of the scattering operator $\sum_{i=1}^{N}\left\{e^{i\,\mathbf{k}_1\mathbf{r}_i} + Ce^{i\,\mathbf{k}_2\mathbf{r}_i}\right\}$ divided by N yields the sum of all Bragg scattering terms in $\mathcal{D}(\mathbf{k}_1,\mathbf{k}_2;C)$, whereas the remaining contribution $N^{-1}\{\cdots\}$ in (55) is the sum of all diffuse scattering terms in $\mathcal{D}(\mathbf{k}_1,\mathbf{k}_2;C)$. Explicitly these relations read

$$V\left\{S_{\mathrm{B}}(0;\mathbf{k}_1) + |C|^2 S_{\mathrm{B}}(0;\mathbf{k}_2) + CS_{\mathrm{B}}(\mathbf{K};\mathbf{k}) + C^* S_{\mathrm{B}}^*(\mathbf{K};\mathbf{k})\right\}$$
$$= \frac{1}{N}\left|\left\langle\sum_{i=1}^{N}\left\{e^{i\,\mathbf{k}_1\mathbf{r}_i} + Ce^{i\,\mathbf{k}_2\mathbf{r}_i}\right\}\right\rangle\right|^2 \tag{56}$$

and

$$S_{\mathrm{d}}(0;\mathbf{k}_1) + |C|^2 S_{\mathrm{d}}(0;\mathbf{k}_2) + CS_{\mathrm{d}}(\mathbf{K};\mathbf{k}) + C^* S_{\mathrm{d}}^*(\mathbf{K};\mathbf{k})$$
$$= \frac{1}{N}\left\{\left\langle\left|\sum_{i=1}^{N}\left\{e^{i\,\mathbf{k}_1\mathbf{r}_i} + Ce^{i\,\mathbf{k}_2\mathbf{r}_i}\right\}\right|^2\right\rangle - \left|\left\langle\sum_{i=1}^{N}\left\{e^{i\,\mathbf{k}_1\mathbf{r}_i} + Ce^{i\,\mathbf{k}_2\mathbf{r}_i}\right\}\right\rangle\right|^2\right\}, \tag{57}$$

where $\mathbf{k}_1$ and $\mathbf{k}_2$ on the one hand and $\mathbf{K}$ and $\mathbf{k}$ on the other are related to each other by means of (44), (45), and (50).

4 Crystallographic Symmetries in the Bragg and Diffuse Scattering Structure Functions

In this section we will discuss the crystallographic symmmetries in the Bragg scattering structure function $S_{\mathrm{B}}(\mathbf{K};\mathbf{k})$, (48), and in the diffuse scattering structure function $S_{\mathrm{d}}(\mathbf{K};\mathbf{k})$, (49).

Let us assume that the crystallographic space group symmetry of the scattering crystal is described by crystallographic space group

$$\mathcal{G} = \big\{\{R, \mathcal{T}_R + \mathbf{t}\} \,\big|\, R \in \mathcal{G}_0;\ \mathbf{t} \text{ a Bravais lattice vector}\big\}\ , \tag{58}$$

consisting of all configuration space operators $\{R, \mathcal{T}_R+\mathbf{t}\}$ for which $\mathbf{t}$ is a vector of the direct Bravais lattice of the crystal and the (proper or improper) rotation R an element of the point group $\mathcal{G}_0$ of crystallographic space group $\mathcal{G}$. The element $\{R, \mathcal{T}_R + \mathbf{t}\} \in \mathcal{G}$ transforms $\mathbf{x} \in |\mathrm{R}^3$ according to

$$\{R, \mathcal{T}_R + \mathbf{t}\}\mathbf{x} = R\mathbf{x} + \mathcal{T}_R + \mathbf{t}\,, \tag{59}$$

where $R\mathbf{x}$ stands for the matrix-vector product of the 3×3 orthogonal matrix R representing rotation R in a Cartesian xyz-frame of reference with the three-dimensional vector $\mathbf{x}$ in this frame. The translation vectors $\mathcal{T}_R$ all vanish for symmorphic space groups $\mathcal{G}$ in their standard settings. For a nonsymmorphic space group (in its standard setting) the $\mathcal{T}_R$ are non-vanishing linear combinations of certain rational multiples of the basis vectors of the direct Bravais lattice for at the least one, in general for several rotations R and vanish for all other rotations R in the point group $\mathcal{G}_0$ of space group $\mathcal{G}$. The $\mathcal{T}_R$ may be found listed in the International Tables for Crystallography, Vol. A [24], for all nonsymmorphic crystallographic space groups in their standard settings.

The point group symmetries in the $\mathbf{K}$-component functions $u(\mathbf{r};\mathbf{K})$ of the two-body density $\varrho_2(\mathbf{S},\mathbf{r})$ are determined by the Laue classes

$$\mathcal{L}_0(\mathbf{K}) = \big\{R', IR' \,\big|\, R' \in \mathcal{G}_0(\mathbf{K});\ R'\mathbf{K} = \mathbf{K}\big\} \tag{60}$$

of the reciprocal lattice vectors $\mathbf{K}$, given by all (proper and improper) rotations R' and IR' for which $R' \in \mathcal{G}_0$ sends $\mathbf{K}$ into itself, i.e. $R'\mathbf{K} = \mathbf{K}$. Here I denotes the inversion through the origin of the Cartesian xyz-frame of reference. The point group

$$\mathcal{G}_0(\mathbf{K}) = \big\{R' \in \mathcal{G}_0 \,\big|\, R'\mathbf{K} = \mathbf{K}\big\} \subseteq \mathcal{G}_0 \tag{61}$$

contains all elements $R' \in \mathcal{G}_0$ that send $\mathbf{K}$ into itself. The Laue class $\mathcal{L}_0(\mathbf{K})$ is obtained from the point group $\mathcal{G}_0(\mathbf{K})$ of reciprocal lattice vector $\mathbf{K}$ by adding the inversion I through the origin.

Component $u(\mathbf{r};\mathbf{K})$ is a basis function of that one-dimensional irreducible representation (IR) $\Gamma_{\tau,\mathcal{L}_0(\mathbf{K})}$ of Laue class $\mathcal{L}_0(\mathbf{K})$ that is given by the mapping

$$\Gamma_{\tau,\mathcal{L}_0(\mathbf{K})}(R') = \Gamma_{\tau,\mathcal{L}_0(\mathbf{K})}(IR') = \mathrm{e}^{i\,\mathbf{K}\mathcal{T}_{R'}}\ , \quad \forall R' \in \mathcal{G}_0(\mathbf{K})\,. \tag{62}$$

This means that $u(\mathbf{r};\mathbf{K})$ transforms under the elements in $\mathcal{L}_0(\mathbf{K})$ according to the prescription

$$\begin{aligned} P_{\mathbf{r}}(R')u(\mathbf{r};\mathbf{K}) &= u\left(R'^{-1}\mathbf{r};\mathbf{K}\right) = P_{\mathbf{r}}(IR')u(\mathbf{r};\mathbf{K}) = u\left((IR')^{-1}\mathbf{r};\mathbf{K}\right) \\ &= u\left(-R'^{-1}\mathbf{r};\mathbf{K}\right) = \Gamma_{\tau,\mathcal{L}_0(\mathbf{K})}(R')u(\mathbf{r};\mathbf{K})\,, \\ &\qquad \forall R' \in \mathcal{G}_0(\mathbf{K})\,, \end{aligned} \tag{63}$$

where ${R'}^{-1} \in \mathcal{G}_0(\mathbf{K}) \subseteq \mathcal{L}_0(\mathbf{K})$ and $(IR')^{-1} = {R'}^{-1}I \in \mathcal{L}_0(\mathbf{K})$ denote the inverse elements of R' and IR', respectively. Note that (63) provides also the definition of how the function space operators $P_\mathbf{r}(R')$ and $P_\mathbf{r}(IR')$ act on functions of variable $\mathbf{r} \in |\mathrm{R}^3$ in the space of functions of variable $\mathbf{r}$.

For symmetry-equivalent reciprocal lattice vectors the lattice Fourier transforms $\rho(\mathbf{K})$ and $u(\mathbf{r};\mathbf{K})$ of the one- and two-body densities are not independent of each other but related by means of

$$\rho(R_j\mathbf{K}) = \mathrm{e}^{-i\,(R_j\mathbf{K})\mathcal{T}_j}\rho(\mathbf{K})\,, \quad \forall j = 1, 2, \cdots, M(\mathbf{K}) \tag{64}$$

and

$$u(\mathbf{r}; R_j\mathbf{K}) = \mathrm{e}^{-i\,(R_j\mathbf{K})\mathcal{T}_j} P_\mathbf{r}(R_j)u(\mathbf{r};\mathbf{K}) = \mathrm{e}^{-i\,(R_j\mathbf{K})\mathcal{T}_j} u\left(R_j^{-1}\mathbf{r};\mathbf{K}\right)\,, \quad \forall j = 1, 2, \cdots, M(\mathbf{K})\,, \tag{65}$$

which equation provides also the definition of how the function space operators $P_\mathbf{r}(R_j)$ act on functions of $\mathbf{r}$. The (proper or improper) $\mathbf{K}$-dependent rotations $R_j \in \mathcal{G}_0$ are $M(\mathbf{K})$ left coset representatives of the decomposition of group $\mathcal{G}_0$ into left cosets w.r.t. subgroup $\mathcal{G}_0(\mathbf{K}) \subseteq \mathcal{G}_0$,

$$\mathcal{G}_0 = \left\{R_jR' \,\middle|\, R' \in \mathcal{G}_0(\mathbf{K});\, 1 \leq j \leq M(\mathbf{K})\right\}\,, \tag{66}$$

where $M(\mathbf{K}) = g_0/g_0(\mathbf{K})$ with g_0 and $g_0(\mathbf{K})$ the orders of (= numbers of elements in) groups $\mathcal{G}_0$ and $\mathcal{G}_0(\mathbf{K})$, respectively. The translation vectors $\mathcal{T}_{R_j}$ (cf. the comments surrounding (59)) associated with the left coset representatives R_j are denoted by $\mathcal{T}_j$, $\mathcal{T}_{R_j} = \mathcal{T}_j$, $1 \leq j \leq M(\mathbf{K})$. Time-reversal symmetry of $\varrho(\mathbf{x})$ and $\varrho_2(\mathbf{S},\mathbf{r})$ (= reality of these quantities) implies

$$\rho(-\mathbf{K}) = \rho^*(\mathbf{K}) \quad \text{and} \quad u(\mathbf{r};-\mathbf{K}) = u^*(\mathbf{r};\mathbf{K})\,, \quad \forall\mathbf{K}\,. \tag{67}$$

The fact that $u(\mathbf{r};\mathbf{K})$ is a basis function of IR $\Gamma_{\tau,\mathcal{L}_0(\mathbf{K})}$ of the Laue class $\mathcal{L}_0(\mathbf{K})$ rather than of IR $\Gamma_{\tau,\mathcal{G}_0(\mathbf{K})}(R') = \mathrm{e}^{i\,\mathbf{K}\mathcal{T}_{R'}}$, $R' \in \mathcal{G}_0(\mathbf{K})$, of just $\mathcal{G}_0(\mathbf{K}) \subseteq \mathcal{L}_0(\mathbf{K})$ derives from the exchange symmetry $u(-\mathbf{r};\mathbf{K}) = u(\mathbf{r};\mathbf{K})$, which relation holds regardless of whether space group $\mathcal{G}$ itself is centrosymmetric or not. We point out that $\rho(\mathbf{K})$ vanishes, if IR $\Gamma_{\tau,\mathcal{L}_0(\mathbf{K})}$ is not the trivial identity IR of $\mathcal{L}_0(\mathbf{K})$, whereas $u(\mathbf{r};\mathbf{K})$ in general does not, even for $\Gamma_{\tau,\mathcal{L}_0(\mathbf{K})}$ not the identity IR of the Laue class $\mathcal{L}_0(\mathbf{K})$ of reciprocal lattice vector $\mathbf{K}$.

The three preceding paragraphs are merely a very succinct summary of crystallographic point and space group symmetries in the one- and two-body densities $\varrho(\mathbf{x})$ and $\varrho_2(\mathbf{S},\mathbf{r})$. These topics are treated at much greater length in full detail with rigorous mathematical proofs in Refs. 1–4.

Invoking (64) one immediately finds

$$S_\mathrm{B}(R_j\mathbf{K};\mathbf{k}) = \mathrm{e}^{-i\,(R_j\mathbf{K})\mathcal{T}_j} P_\mathbf{k}(R_j)S_\mathrm{B}(\mathbf{K};\mathbf{k}) = \mathrm{e}^{-i\,(R_j\mathbf{K})\mathcal{T}_j} S_\mathrm{B}\left(\mathbf{K}; R_j^{-1}\mathbf{k}\right)\,, \quad \forall j = 1, 2, \cdots, M(\mathbf{K})\,, \tag{68}$$

where the function space operator $P_\mathbf{k}(R_j)$ acts on variable $\mathbf{k}$ as defined in this equation. Equation (68) relates those Bragg scattering structure factors, defined in (48), that belong to symmetry-equivalent reciprocal lattice vectors $R_j\mathbf{K}$, $1 \leq j \leq M(\mathbf{K})$.

By making use of $\rho(R\mathbf{K}) = \mathrm{e}^{-i(R\mathbf{K})\mathcal{T}_R}\rho(\mathbf{K})$, which relation holds for all elements $R \in \mathcal{G}_0$, and of time-reversal symmetry $\rho(-\mathbf{K}) = \rho^*(\mathbf{K})$ one immediately derives that, as functions of variable $\mathbf{k}$, the Bragg scattering structure functions $S_{\mathrm{B}}(\mathbf{K};\mathbf{k})$ are basis functions of the one-dimensional IRs $\Gamma_{\tau,\mathcal{L}_0(\mathbf{K})}$, (62), of the Laue classes $\mathcal{L}_0(\mathbf{K})$, (60), i.e.

$$P_{\mathbf{k}}(R')S_{\mathrm{B}}(\mathbf{K};\mathbf{k}) = S_{\mathrm{B}}\left(\mathbf{K};R'^{-1}\mathbf{k}\right) = \Gamma_{\tau,\mathcal{L}_0(\mathbf{K})}(R')S_{\mathrm{B}}(\mathbf{K};\mathbf{k})\,, \quad \forall R' \in \mathcal{L}_0(\mathbf{K})\,, \tag{69}$$

where again the action of the function space operators $P_{\mathbf{k}}(R')$ is evident. In proving the transformation properties (69) of $S_{\mathrm{B}}(\mathbf{K};\mathbf{k})$ under the elements of $\mathcal{L}_0(\mathbf{K})$ one has to make use also of $\mathcal{T}_{R^{-1}} = -R^{-1}\mathcal{T}_R + \mathbf{t}_{R^{-1}}$, where $\mathbf{t}_{R^{-1}}$ is a Bravais lattice vector. We mention that $S_{\mathrm{B}}(\mathbf{K};\mathbf{k})$ displays symmetry-imposed systematic extinctions, whenever $\rho\left(\mathbf{k}+\frac{1}{2}\mathbf{K}\right)$ and/or $\rho\left(\mathbf{k}-\frac{1}{2}\mathbf{K}\right)$ vanishes for symmetry reasons. Endowed with operator multiplicaton $P_{\mathbf{k}}(R'_1)P_{\mathbf{k}}(R'_2) = P_{\mathbf{k}}(R'_1R'_2)$ for all R'_1, $R'_2 \in \mathcal{L}_0(\mathbf{K})$ the function space operators $\left\{P_{\mathbf{k}}(R') \,\middle|\, R' \in \mathcal{L}_0(\mathbf{K})\right\}$ constitute a realization of group $\mathcal{L}_0(\mathbf{K})$ in the space of functions of variable $\mathbf{k} \in \mathbb{R}^3$.

Exploiting (64), (65), and (68) one derives

$$S_{\mathrm{d}}(R_j\mathbf{K};\mathbf{k}) = \mathrm{e}^{-i(R_j\mathbf{K})\mathcal{T}_j}P_{\mathbf{k}}(R_j)S_{\mathrm{d}}(\mathbf{K};\mathbf{k}) = \mathrm{e}^{-i(R_j\mathbf{K})\mathcal{T}_j}S_{\mathrm{d}}\left(\mathbf{K};R_j^{-1}\mathbf{k}\right)\,, \quad \forall j = 1,2,\cdots,M(\mathbf{K})\,, \tag{70}$$

relating those diffuse background scattering factors, defined in (49), for $\mathbf{K}$-vectors that are symmetry-equivalent.

Invoking (63) and (69) one arrives at

$$P_{\mathbf{k}}(R')S_{\mathrm{d}}(\mathbf{K};\mathbf{k}) = S_{\mathrm{d}}\left(\mathbf{K};R'^{-1}\mathbf{k}\right) = \Gamma_{\tau,\mathcal{L}_0(\mathbf{K})}(R')S_{\mathrm{d}}(\mathbf{K};\mathbf{k})\,, \quad \forall R' \in \mathcal{L}_0(\mathbf{K})\,, \tag{71}$$

showing that also the diffuse scattering functions $S_{\mathrm{d}}(\mathbf{K};\mathbf{k})$ transform as basis functions of the one-dimensional IRs $\Gamma_{\tau,\mathcal{L}_0(\mathbf{K})}$, (62), of the Laue classes $\mathcal{L}_0(\mathbf{K})$, (60). In proving (71) one has to bear in mind that $\rho(\mathbf{K}) = 0$ for IR $\Gamma_{\tau,\mathcal{L}_0(\mathbf{K})}$ not the identity IR of $\mathcal{L}_0(\mathbf{K})$.

Time-reversal symmetry for $S_{\mathrm{B}}(\mathbf{K};\mathbf{k})$ is a trivial consequence of the very definition (48) of $S_{\mathrm{B}}(\mathbf{K};\mathbf{k})$ and simply reads

$$S_{\mathrm{B}}(-\mathbf{K};\mathbf{k}) = S_{\mathrm{B}}^*(\mathbf{K};\mathbf{k})\,, \tag{72}$$

which result in conjunction with $\rho(-\mathbf{K}) = \rho^*(\mathbf{K})$ and $u(\mathbf{r};-\mathbf{K}) = u^*(\mathbf{r};\mathbf{K})$ yields the time-reversal symmetry

$$S_{\mathrm{d}}(-\mathbf{K};\mathbf{k}) = S_{\mathrm{d}}^*(\mathbf{K};\mathbf{k}) \tag{73}$$

of the diffuse scattering function $S_{\mathrm{d}}(\mathbf{K};\mathbf{k})$, (49). The transformation properties (69) and (71) of the Bragg and diffuse scattering structure functions $S_{\mathrm{B}}(\mathbf{K};\mathbf{k})$ and $S_{\mathrm{d}}(\mathbf{K};\mathbf{k})$ in the space of functions of variable $\mathbf{k}$ are analoguous to the transformation properties (63) of the lattice Fourier transform functions $u(\mathbf{r};\mathbf{K})$ of the full two-body density $\varrho_2(\mathbf{S},\mathbf{r})$ in the space of functions of variable $\mathbf{r}$. Likewise, (68) and (70) relate the Bragg and diffuse scattering structure factors $S_{\mathrm{B}}(\mathbf{K};\mathbf{k})$ and $S_{\mathrm{d}}(\mathbf{K};\mathbf{k})$ for symmetry-equivalent reciprocal lattice vectors $R_j\mathbf{K}$, $1 \le j \le M(\mathbf{K})$, to each other in a manner analoguous to how the lattice Fourier coefficients $\rho(\mathbf{K})$ of the one-body density $\varrho(\mathbf{x})$ and the two-body density Fourier coefficient functions $u(\mathbf{r};\mathbf{K})$ are related to each other for symmetry-equivalent reciprocal lattice vectors by means of (64) and (65), respectively.

We conclude by drawing attention to a few important points in the above findings.

For $\mathbf{K}=0$ the corresponding Laue class $\mathcal{L}_0(0)=\mathcal{L}_0$ is the Laue class $\mathcal{L}_0$ to which crystallographic space group $\mathcal{G}$ and its associated crystallographic point group $\mathcal{G}_0=\mathcal{G}_0(0)$ (cf. (61)) belong. Furthermore the IR $\Gamma_{\tau,\mathcal{L}_0(0)}=\Gamma_{\tau,\mathcal{L}_0}$, (62), is just the trivial identity representation of $\mathcal{L}_0(0)$ for $\mathbf{K}=0$, i.e. $\Gamma_{\tau,\mathcal{L}_0(0)}(R')=1$, $\forall R' \in \mathcal{L}_0(0)$. For $\mathbf{K}=0$ the Bragg scattering structure factor $S_{\mathrm{B}}(0;\mathbf{k})$ obviously is real. Because of this and because $u(\mathbf{r};0)$ is real and centrosymmetric, also the diffuse scattering structure factor $S_{\mathrm{d}}(0;\mathbf{k})$, for reciprocal lattice vector $\mathbf{K}=0$, is real. We furthermore note that (69) and (71) imply that both $S_{\mathrm{B}}(\mathbf{K};\mathbf{k})$ and $S_{\mathrm{d}}(\mathbf{K};\mathbf{k})$ are centrosymmetric functions of $\mathbf{k}$ for all reciprocal lattice vectors $\mathbf{K}$. The analysis carried out above shows that both the Bragg and diffuse scattering structure functions $S_{\mathrm{B}}(\mathbf{K};\mathbf{k})$ and $S_{\mathrm{d}}(\mathbf{K};\mathbf{k})$ follow exactly the same crystallographic symmetries as the $\mathbf{K}$-components $u(\mathbf{r};\mathbf{K})$ of the two-body density $\varrho_2(\mathbf{S},\mathbf{r})$. The $\mathbf{K}=0$ diffuse scattering structure factor $S_{\mathrm{d}}(0;\mathbf{k})$ can provide information only about the Laue class $\mathcal{L}_0$ to which the space group $\mathcal{G}$ of the scattering material belongs, whereas information about the translational symmetries in $\mathcal{G}$ and about the IRs $\Gamma_{\tau,\mathcal{L}_0(\mathbf{K})}$ of the Laue classes $\mathcal{L}_0(\mathbf{K})$ of the $\mathbf{K}$-vectors must come from $\mathbf{K}\neq 0$ scattering. We furthermore point out that even where the Bragg scattering function $S_{\mathrm{B}}(\mathbf{K};\mathbf{k})$ exhibits symmetry-caused systematic extinctions, the diffuse scattering function $S_{\mathrm{d}}(\mathbf{K};\mathbf{k})$ in general does not vanish. In general there will be a diffuse background in the scattering cross section. The comments in this paragraph apply to any crystallographic space group $\mathcal{G}$.

5 Bragg and Diffuse Scattering in X-Ray Diffraction of a Double-Beam from Condensed Matter

To relate the Fourier transform $\rho(\mathbf{K})$, (38), of the atomic one-body density $\varrho(\mathbf{x})$, (39), and the Fourier transform function $u(\mathbf{r};\mathbf{K})$, (42), of the atomic two-body density, (40), to experimentally measurable X-ray diffraction intensities, we first observe that the scattering cross section $\left(\frac{d\Omega}{d\sigma}\right)_{\mathrm{X}}$ of X-ray diffraction is related to the canonical density matrix of the electrons by means of the proportionality relation [25]

$$\left(\frac{d\Omega}{d\sigma}\right)_{\mathrm{X}} \sim \left\langle \left| \sum_{k=1}^{M} \left\{ \mathrm{e}^{i\,\mathbf{k}_1\mathbf{r}_k} + C\mathrm{e}^{i\,\mathbf{k}_2\mathbf{r}_k} \right\} \right|^2 \right\rangle_{\mathrm{e}} . \tag{74}$$

The sum on the left side of this proportionality relation extends over all electrons, with position vectors $\mathbf{r}_k$, $1 \leq k \leq M$, of the scattering many-body system. The total number M of electrons is given by $M = NZ$, where N is the total number of atoms of the scatterer and Z the total number of electrons in an atom of the scatterer. As above we consider X-ray diffraction only from a crystal made out of identical atoms. The Dirac brackets $\langle \cdots \rangle_{\mathrm{e}}$ in relation (74) denote the thermodynamic expectation value of the absolute square of the many-electron scattering operator $\sum_{k=1}^{M} \{\mathrm{e}^{i\,\mathbf{k}_1 \mathbf{r}_k} + C\mathrm{e}^{i\,\mathbf{k}_2 \mathbf{r}_k}\}$ with respect to the canonical density operator of the many-electron system formed by the electrons of the scatterer's atoms (cf. (55)). As before we consider the diffraction of an incoming X-ray beam that is the linear superposition of two plane waves of the same energy. The changes in the wave number vectors (cf. (29)) of the incident X-ray double-beam as a result of the diffraction process are denoted by $\mathbf{k}_1$ and $\mathbf{k}_2$ in (74). In contrast to neutrons, which are scattered from the atomic nuclei of the scattering many-body system, X-rays are diffracted by the electrons of the scatterer.

In complete analogy to (55) and (36) it may now be shown, after some rather straightforward algebra, that relation (74) may be cast in the form

$$
\begin{aligned}
\left(\frac{d\Omega}{d\sigma}\right)_{\mathrm{X}} \sim M &+ \int \varrho_2^{(\mathrm{e})}(\mathbf{x}_1, \mathbf{x}_2)\mathrm{e}^{-i\,\mathbf{k}_1(\mathbf{x}_1-\mathbf{x}_2)}\, d\mathbf{x}_1 d\mathbf{x}_2 \\
&+ |C|^2 \left\{ M + \int \varrho_2^{(\mathrm{e})}(\mathbf{x}_1, \mathbf{x}_2)\mathrm{e}^{-i\,\mathbf{k}_2(\mathbf{x}_1-\mathbf{x}_2)}\, d\mathbf{x}_1 d\mathbf{x}_2 \right\} \\
&+ C \left\{ \int \varrho_{\mathrm{e}}(\mathbf{x})\mathrm{e}^{-i\,(\mathbf{k}_1-\mathbf{k}_2)\mathbf{x}}\, d\mathbf{x} \right. \\
&\qquad \left. + \int \varrho_2^{(\mathrm{e})}(\mathbf{x}_1, \mathbf{x}_2)\mathrm{e}^{-i\,(\mathbf{k}_1\mathbf{x}_1-\mathbf{k}_2\mathbf{x}_2)}\, d\mathbf{x}_1 d\mathbf{x}_2 \right\} \\
&+ C^* \left\{ \int \varrho_{\mathrm{e}}(\mathbf{x})\mathrm{e}^{-i\,(\mathbf{k}_1-\mathbf{k}_2)\mathbf{x}}\, d\mathbf{x} \right. \\
&\qquad \left. + \int \varrho_2^{(\mathrm{e})}(\mathbf{x}_1, \mathbf{x}_2)\mathrm{e}^{-i\,(\mathbf{k}_1\mathbf{x}_1-\mathbf{k}_2\mathbf{x}_2)}\, d\mathbf{x}_1 d\mathbf{x}_2 \right\}^* ,
\end{aligned}
\tag{75}
$$

wherein $\varrho_{\mathrm{e}}(\mathbf{x})$ is the electronic one-body density and $\varrho_2^{(\mathrm{e})}(\mathbf{x}_1, \mathbf{x}_2)$ the electronic two-body density of the scattering material.

To express the proportionality relation (75) in terms of the atomic one- and two-body densities $\varrho(\mathbf{x})$ and $\rho_2(\mathbf{x}_1, \mathbf{x}_2)$, introduced further above in (39) and (37), we consider the one-electron density operator

$$
\hat{\varrho}_{\mathrm{e}}(\mathbf{x}) = \sum_{i=1}^{N} \sum_{\kappa=1}^{Z} \delta(\mathbf{x} - \mathbf{r}_\kappa - \mathbf{R}_i) \, , \tag{76}
$$

wherein $\mathbf{R}_i \in |\mathrm{R}^3$ denotes the position of the center of atom i, $1 \leq i \leq N$, of the scatterer and $\mathbf{r}_\kappa \in |\mathrm{R}^3$, $1 \leq \kappa \leq Z$, the position of electron κ relative to the center of the atom at $\mathbf{R}_i$. Relative to the center-of-atom position $\mathbf{R}_i$ the one-electron density $n_{\mathrm{e}}(\mathbf{x} - \mathbf{R}_i)$ of atom i may be written as

$$n_{\mathrm{e}}(\mathbf{x}-\mathbf{R}_i)=\sum_{\kappa=1}^{Z}\delta(\mathbf{x}-\mathbf{R}_i-\mathbf{r}_\kappa)\,. \tag{77}$$

The one-electron density $\varrho_{\mathrm{e}}(\mathbf{x})$ of the total many-electron system of the diffracting material may then be obtained as the sum of the contributions (77) of all N particles to $\varrho_{\mathrm{e}}(\mathbf{x})$, yielding

$$\begin{aligned}\varrho_{\mathrm{e}}(\mathbf{x}) &= \sum_{i=1}^{N} n_{\mathrm{e}}(\mathbf{x}-\mathbf{R}_i)=\sum_{i=1}^{N}\int n_{\mathrm{e}}(\mathbf{x}')\delta(\mathbf{x}-\mathbf{R}_i-\mathbf{x}')\,d\mathbf{x}'\\ &= \int n_{\mathrm{e}}(\mathbf{x}')\left\{\sum_{i=1}^{N}\delta(\mathbf{x}-\mathbf{x}'-\mathbf{R}_i)\right\}d\mathbf{x}'=\int n_{\mathrm{e}}(\mathbf{x}')\varrho(\mathbf{x}-\mathbf{x}')\,d\mathbf{x}'\,,\end{aligned} \tag{78}$$

where $\varrho(\mathbf{x}-\mathbf{x}')=\sum_{i=1}^{N}\delta(\mathbf{x}-\mathbf{x}'-\mathbf{R}_i)$ is the atomic one-body density at $\mathbf{x}-\mathbf{x}'$. It is straightforward to prove that (78) may be also written as

$$\varrho_{\mathrm{e}}(\mathbf{x})=\int \varrho(\mathbf{x}')n_{\mathrm{e}}(\mathbf{x}-\mathbf{x}')\,d\mathbf{x}'\,, \tag{79}$$

rendering the local one-electron density $\varrho_{\mathrm{e}}(\mathbf{x})$ of the scatterer in the form of the convolution integral of the atomic local particle number density $\varrho(\mathbf{x}')$ with the electron distribution function $n_{\mathrm{e}}(\mathbf{x}-\mathbf{x}')$ describing the distribution of the Z electrons in an atom centered at $\mathbf{x}'$. The three-dimensional integrations $\int\cdots d\mathbf{x}'$ in (78) and (79) extend over all of three-dimensional space $|\mathrm{R}^3$. Equation (79) is a well-known and widely employed approximation for the local electron density $\varrho_{\mathrm{e}}(\mathbf{x})$. The form (79) for $\varrho_{\mathrm{e}}(\mathbf{x})$ shows that here it is assumed that the distribution $n_{\mathrm{e}}(\mathbf{x}-\mathbf{x}')$ of electrons about an atom center at $\mathbf{x}'$ does not change due to the presence of other atoms in the scattering material and that the motion of the electrons in the atoms follows the motion of the centers of the atoms in the scattering many-body system, for which reason formula (79) may be referred to as an adiabatic approximation.

Equation (79) implies in particular that the structure factor

$$\mathcal{S}(\mathbf{K}')=\int_{\Omega}\varrho_{\mathrm{e}}(\mathbf{x})\mathrm{e}^{-i\mathbf{K}'\mathbf{x}}\,d\mathbf{x}=\Omega\rho(\mathbf{K}')\mathcal{F}(K')\,, \tag{80}$$

where $K'=|\mathbf{K}'|$, of a crystalline material assumes the form of the product of $\Omega\rho(\mathbf{K}')$ with the atomic form factor

$$\mathcal{F}(K')=\int n_{\mathrm{e}}(\mathbf{x})\mathrm{e}^{-i\mathbf{K}'\mathbf{x}}\,d\mathbf{x}\,, \tag{81}$$

the Fourier transform of the atomic electron distribution function $n_{\mathrm{e}}(\mathbf{x})$. The structure factor $\mathcal{S}(\mathbf{K}')$ defined in (80) is the lattice Fourier transform of the one-electron density $\varrho_{\mathrm{e}}(\mathbf{x})$ of the crystal. The integration in (80) extends over the volume Ω of a primitive unit cell of the crystal, whereas the integral in the above given definition (81) of the atomic form factor $\mathcal{F}(K')$ is over all of three-dimensional space $|\mathrm{R}^3$. Obviously $\mathcal{S}(\mathbf{K}')$ vanishes for $\mathbf{K}'$ not a reciprocal lattice vector of the crystalline material. It also vanishes for those reciprocal lattice vectors $\mathbf{K}'$ for which $\rho(\mathbf{K}')$ vanishes, in particular for those $\mathbf{K}'$ for which the lattice Fourier transform $\rho(\mathbf{K}')$ of the atomic one-body density $\varrho(\mathbf{x})$ of the crystal displays symmetry-caused systematic extinctions, i.e. for those $\mathbf{K}'$ for which the one-dimensional irreducible representations $\Gamma_{\tau,\mathcal{L}_0(\mathbf{K}')}$ in (62) of the Laue classes $\mathcal{L}_0(\mathbf{K}')$, defined in (60), are not the trivial identity representations of the (crystallographic point) groups $\mathcal{L}_0(\mathbf{K}')$. The atomic form factor $\mathcal{F}(K')$ in (80) depends on only the magnitude K' for spherically symmetric atomic electron distributions $n_e(\mathbf{x})$, i.e. $n_e(\mathbf{x}) = n_e(|\mathbf{x}|)$ depends on only the distance $|\mathbf{x}|$ from the center of the atom. In the present work we consider only systems for which this assumption is valid, at the least to an acceptable degree of accuracy. For reciprocal lattice vector $\mathbf{K}' = 0$ formula (80) yields for the associated $\mathbf{K}' = 0$ structure factor $\mathcal{S}(0)$ the trivial result $\mathcal{S}(0) = N_\Omega Z$, which is just the total number of electrons in a primitive unit cell. Here N_Ω is the number of atoms in a primitive unit cell and Z as above the number of electrons in an atom of the scatterer. Because the atomic form factor $\mathcal{F}(K')$ in (80) is assumed to be a spherically symmetric function of its argument $\mathbf{K}'$, the structure factor $\mathcal{S}(\mathbf{K}')$ follows exactly the same crystallographic symmetries as the lattice Fourier transform $\rho(\mathbf{K}')$ of the local one-body density $\varrho(\mathbf{x})$. This applies in particular to (64). The structure factors $\mathcal{S}(R_j\mathbf{K}')$ for symmetry-equivalent reciprocal lattice vectors $R_j\mathbf{K}'$, $1 \le j \le M(\mathbf{K}')$, are related to each other by means of the same transformations as in (64).

In analogy to the double sum $\sum_{i\neq j=1}^{N} \delta(\mathbf{x}_1 - \mathbf{r}_i)\delta(\mathbf{x}_2 - \mathbf{r}_j)$ in the definition (37) of the atomic two-body density $\rho_2(\mathbf{x}_1, \mathbf{x}_2)$ the two-electron density operator $\hat{\rho}_2^{(e)}(\mathbf{x}_1, \mathbf{x}_2)$ may be cast in the form

$$\begin{aligned}\hat{\rho}_2^{(e)}(\mathbf{x}_1, \mathbf{x}_2) &= \sum_{i=1}^{N}\sum_{j=1}^{N}\sum_{\kappa=1}^{Z}\sum_{\ell=1}^{Z} \delta(\mathbf{x}_1 - \mathbf{r}_\kappa - \mathbf{R}_i)\delta(\mathbf{x}_2 - \mathbf{r}_\ell - \mathbf{R}_j) \\ &\quad - \sum_{i=1}^{N}\sum_{\kappa=1}^{Z} \delta(\mathbf{x}_1 - \mathbf{r}_\kappa - \mathbf{R}_i)\delta(\mathbf{x}_2 - \mathbf{r}_\kappa - \mathbf{R}_i)\,,\end{aligned} \tag{82}$$

where, similarly to (76), $\mathbf{r}_\kappa \in |\mathrm{R}^3$ and $\mathbf{r}_\ell \in |\mathrm{R}^3$, $1 \le \kappa, \ell \le Z$, denote the electron positions relative to the center of an atom and $\mathbf{R}_i \in |\mathrm{R}^3$ and $\mathbf{R}_j \in |\mathrm{R}^3$, $1 \le i, j \le N$, the positions of the centers of the atoms. Subtracting the second expression on the right side of (82) from the first ensures that pairs of same electrons are excluded from the double sum over all pairs of electrons, i.e. the summation in (82) extends only over all pairs of different electrons (cf. (37)).

Obvious rearrangements in the expression (82) yield the two-electron density operator in the form

$$\begin{aligned}\hat{\rho}_2^{(e)}(\mathbf{x}_1, \mathbf{x}_2) &= \sum_{i\neq j=1}^{N}\sum_{\kappa=1}^{Z}\sum_{\ell=1}^{Z} \delta(\mathbf{x}_1 - \mathbf{r}_\kappa - \mathbf{R}_i)\delta(\mathbf{x}_2 - \mathbf{r}_\ell - \mathbf{R}_j) \\ &\quad + \sum_{i=1}^{N}\sum_{\kappa\neq\ell=1}^{Z} \delta(\mathbf{x}_1 - \mathbf{r}_\kappa - \mathbf{R}_i)\delta(\mathbf{x}_2 - \mathbf{r}_\ell - \mathbf{R}_i)\,.\end{aligned} \tag{83}$$

Observing that the double sum $\sum_{\kappa\neq\ell=1}^{Z}\cdots$ in the second term on the right side of (83) is the two-electron density $n_{\rm e}^{(2)}(\mathbf{x}_1-\mathbf{R}_i,\mathbf{x}_2-\mathbf{R}_i)$, relative to the center $\mathbf{R}_i$ of atom i, of the Z electrons in atom i at position $\mathbf{R}_i$,

$$n_{\rm e}^{(2)}(\mathbf{x}_1-\mathbf{R}_i,\mathbf{x}_2-\mathbf{R}_i)=\sum_{\kappa\neq\ell=1}^{Z}\delta(\mathbf{x}_1-\mathbf{r}_\kappa-\mathbf{R}_i)\delta(\mathbf{x}_2-\mathbf{r}_\ell-\mathbf{R}_i)\,, \tag{84}$$

and employing in the first double sum $\sum_{i\neq j=1}^{N}\cdots$ on the right of (83) the form (77) for the electron distribution of atom i, the two-electron density $\rho_2^{(\rm e)}(\mathbf{x}_1,\mathbf{x}_2)$ may be written as

$$\rho_2^{(\rm e)}(\mathbf{x}_1,\mathbf{x}_2)=\sum_{i\neq j=1}^{N}n_{\rm e}(\mathbf{x}_1-\mathbf{R}_i)n_{\rm e}(\mathbf{x}_2-\mathbf{R}_i)+\sum_{i=1}^{N}n_{\rm e}^{(2)}(\mathbf{x}_1-\mathbf{R}_i,\mathbf{x}_2-\mathbf{R}_i)\,. \tag{85}$$

Equation (85) may be recast to yield

$$\begin{aligned}\rho_2^{(\rm e)}(\mathbf{x}_1,\mathbf{x}_2) &= \int n_{\rm e}(\mathbf{x}_1')n_{\rm e}(\mathbf{x}_2')\\ &\quad\times\left\{\sum_{i\neq j=1}^{N}\delta(\mathbf{x}_1-\mathbf{x}_1'-\mathbf{R}_i)\delta(\mathbf{x}_2-\mathbf{x}_2'-\mathbf{R}_j)\right\}d\mathbf{x}_1'd\mathbf{x}_2'\\ &\quad+\int n_{\rm e}^{(2)}(\mathbf{x}_1-\mathbf{x}',\mathbf{x}_2-\mathbf{x}')\left\{\sum_{i=1}^{N}\delta(\mathbf{x}'-\mathbf{R}_i)\right\}d\mathbf{x}'\,.\end{aligned} \tag{86}$$

Since $\sum_{i\neq j=1}^{N}\delta(\mathbf{x}_1-\mathbf{x}_1'-\mathbf{R}_i)\delta(\mathbf{x}_2-\mathbf{x}_2'-\mathbf{R}_j)$ is the atomic two-body density $\rho_2(\mathbf{x}_1-\mathbf{x}_1',\mathbf{x}_2-\mathbf{x}_2')$ of the crystal and $\sum_{i=1}^{N}\delta(\mathbf{x}'-\mathbf{R}_i)$ its atomic one-body density $\varrho(\mathbf{x}')$, one obtains from (86) the result

$$\begin{aligned}\rho_2^{(\rm e)}(\mathbf{x}_1,\mathbf{x}_2) &= \int n_{\rm e}(\mathbf{x}_1')n_{\rm e}(\mathbf{x}_2')\rho_2(\mathbf{x}_1-\mathbf{x}_1',\mathbf{x}_2-\mathbf{x}_2')\,d\mathbf{x}_1'd\mathbf{x}_2'\\ &\quad+\int n_{\rm e}^{(2)}(\mathbf{x}_1-\mathbf{x}',\mathbf{x}_2-\mathbf{x}')\varrho(\mathbf{x}')\,d\mathbf{x}'\end{aligned} \tag{87}$$

for the two-electron density $\rho_2^{(\rm e)}(\mathbf{x}_1,\mathbf{x}_2)$ of the diffracting many-body system. Formula (87) for $\rho_2^{(\rm e)}(\mathbf{x}_1,\mathbf{x}_2)$ is the generalization of (78) to the two-electron density in the adiabatic approximation, in which it is assumed that the one- and two-electron distribution functions $n_{\rm e}(\mathbf{x}_1')$, $n_{\rm e}(\mathbf{x}_2')$, and $n_{\rm e}^{(2)}(\mathbf{x}_1-\mathbf{x}',\mathbf{x}_2-\mathbf{x}')$ of the Z electrons in an atom are not altered by the presence of other atoms and that the electrons follow the motion of the nuclei of the particles in the scatterer. Equation (87) relates $\rho_2^{(\rm e)}(\mathbf{x}_1,\mathbf{x}_2)$ to the atomic one- and two-body densities by means of the generalized convolution integrals on the right of this equation.

To proceed from (87), we now use a further approximation of the form

$$n_{\rm e}^{(2)}(\mathbf{x}_1',\mathbf{x}_2')=n_{\rm e}(\mathbf{x}_1')n_{\rm e}(\mathbf{x}_2')-n_{\rm e}(\mathbf{x}_2')\delta(\mathbf{x}_1'-\mathbf{x}_2') \tag{88}$$

for the two-electron distribution function $n_\mathrm{e}^{(2)}(\mathbf{x}_1', \mathbf{x}_2')$ of the Z electrons in an atom. The approximation (88) is inspired by the fact that $n_\mathrm{e}^{(2)}(\mathbf{x}_1', \mathbf{x}_2')$ approaches the product $n_\mathrm{e}(\mathbf{x}_1')n_\mathrm{e}(\mathbf{x}_2')$ in the limit of large electron distances $|\mathbf{x}_1' - \mathbf{x}_2'| \longrightarrow \infty$ (cf. (52)) and that $n_\mathrm{e}^{(2)}(\mathbf{x}_1', \mathbf{x}_2') = \left(1 - Z^{-1}\right) n_\mathrm{e}(\mathbf{x}_1')n_\mathrm{e}(\mathbf{x}_2')$ for Z non-interacting electrons. Subtracting in (88) the term $n_\mathrm{e}(\mathbf{x}_1')\delta(\mathbf{x}_1' - \mathbf{x}_2')$ from the product $n_\mathrm{e}(\mathbf{x}_1')n_\mathrm{e}(\mathbf{x}_2')$ ensures that no two electrons are simultaneously at the same position in three-dimensional space $|\mathrm{R}^3$, which situation would be unphysical, and also ensures the correct normalization $\int n_\mathrm{e}^{(2)}(\mathbf{x}_1', \mathbf{x}_2')\, d\mathbf{x}_1' d\mathbf{x}_2' = Z(Z-1)$ of the approximation (88) for the two-electron distribution function $n_\mathrm{e}^{(2)}(\mathbf{x}_1', \mathbf{x}_2')$ of the Z electrons in an atom.

Inserting the approximation (88) in (87) finally yields

$$\begin{aligned} \rho_2^{(\mathrm{e})}(\mathbf{x}_1, \mathbf{x}_2) &= \int n_\mathrm{e}(\mathbf{x}_1')n_\mathrm{e}(\mathbf{x}_2')\rho_2(\mathbf{x}_1 - \mathbf{x}_1', \mathbf{x}_2 - \mathbf{x}_2')\, d\mathbf{x}_1' d\mathbf{x}_2' \\ &\quad + \int n_\mathrm{e}(\mathbf{x}_1 - \mathbf{x}')n_\mathrm{e}(\mathbf{x}_2 - \mathbf{x}')\varrho(\mathbf{x}')\, d\mathbf{x}' \\ &\quad -\delta(\mathbf{x}_1 - \mathbf{x}_2)\varrho_\mathrm{e}(\mathbf{x}_1)\,. \end{aligned} \tag{89}$$

Equations (78) and (89) relate the one- and two-electron densities $\varrho_\mathrm{e}(\mathbf{x})$ and $\rho_2^{(\mathrm{e})}(\mathbf{x}_1, \mathbf{x}_2)$ of the X-ray diffracting material to the atomic one- and two-body densities $\varrho(\mathbf{x})$ and $\rho_2(\mathbf{x}_1, \mathbf{x}_2)$ via the electronic density distribution function $n_\mathrm{e}(\mathbf{x})$ of the Z electrons in an atom, which is known, because the atomic form factors $\mathcal{F}(K)$ are known. The atomic form factors $\mathcal{F}(K)$ are specific for the scattering material. Numerical values for the atomic form factors may be taken from the International Tables for Crystallography, Vol. C [23].

Inserting now the results (78) (last expression in this equation) and (89) for $\varrho_\mathrm{e}(\mathbf{x})$ and $\rho_2^{(\mathrm{e})}(\mathbf{x}_1, \mathbf{x}_2)$ in the proportionality relation (75) and making repeatedly use of the convolution theorem, one finds after some lengthy but else rather straightforward algebra

$$\begin{aligned} \frac{1}{N}\left(\frac{d\Omega}{d\sigma}\right)_\mathrm{X} &\sim |\mathcal{F}(k_1)|^2 \left\{V S_\mathrm{B}(0;\mathbf{k}_1) + S_\mathrm{d}(0;\mathbf{k}_1)\right\} \\ &\quad + |C|^2 |\mathcal{F}(k_2)|^2 \left\{V S_\mathrm{B}(0;\mathbf{k}_2) + S_\mathrm{d}(0;\mathbf{k}_2)\right\} \\ &\quad + C\mathcal{F}(k_1)\mathcal{F}^*(k_2) \left\{V S_\mathrm{B}(\mathbf{K};\mathbf{k}) + S_\mathrm{d}(\mathbf{K};\mathbf{k})\right\} \\ &\quad + C^*\mathcal{F}^*(k_1)\mathcal{F}(k_2) \left\{V S_\mathrm{B}^*(\mathbf{K};\mathbf{k}) + S_\mathrm{d}^*(\mathbf{K};\mathbf{k})\right\} \end{aligned} \tag{90}$$

for the total differential cross section $\left(\frac{d\Omega}{d\sigma}\right)_\mathrm{X}$ of X-ray diffraction of an incident double-beam. Here $S_\mathrm{B}(\mathbf{K};\mathbf{k})$ and $S_\mathrm{d}(\mathbf{K};\mathbf{k})$ are the Bragg and diffuse scattering structure functions introduced in (48) and (49), $\mathbf{K}$ and $\mathbf{k}$ are as in (44) and (45), and $k_1 = |\mathbf{k}_1|$ and $k_2 = |\mathbf{k}_2|$. Relation (90) shows that also the differential cross section of X-ray diffraction consists of volume-extensive dominant Bragg peaks and diffuse background scattering terms. As before the dominant volume-extensive Bragg scattering peaks originate in the asymptotic long-range behavior (52) of the two-body density. As before the difference $\mathbf{K}$ of the two wave numbers in an incoming double-beam must be a reciprocal lattice vector of the diffracting crystal for $S_{\mathrm{d}}(\mathbf{K};\mathbf{k})$ to be non-zero. The diffuse scattering of an incident genuinely double, i.e. $\mathbf{K}\neq 0$, beam would lead to information about the $\mathbf{K}\neq 0$ Fourier coefficient functions $u(\mathbf{r};\mathbf{K})$ in the lattice Fourier series representation (40) of the full two-body density $\varrho_2(\mathbf{S},\mathbf{r})$ of the diffracting material. In complete analogy to formula (47) for the total scattering structure function $\mathcal{D}(\mathbf{k}_1,\mathbf{k}_2;C)$, discussed in Sect. 3 in the context of scattering of neutrons from crystalline matter, the differential cross section $(\frac{d\Omega}{d\sigma})_{\mathrm{X}}$ of X-ray diffraction in the proportionality relation (90) splits up into terms that depend on only either $\mathbf{k}_1$ or $\mathbf{k}_2$ and into terms that depend on both momenta transfers $\mathbf{k}_1$ and $\mathbf{k}_2$ from the diffracted X-ray double-beam to the crystalline scatterer or, equivalently, on both $\mathbf{K}$ and $\mathbf{k}$ (cf. (44) and (45)). The former two contributions to the X-ray diffraction differential cross section $(\frac{d\Omega}{d\sigma})_{\mathrm{X}}$ arise from the scattering of the two plane waves $\mathrm{e}^{i\,\mathbf{k}_1^{(\mathrm{in})}\mathbf{x}}$ and $\mathrm{e}^{i\,\mathbf{k}_2^{(\mathrm{in})}\mathbf{x}}$ in the incoming two-beam $\mathrm{e}^{i\,\mathbf{k}_1^{(\mathrm{in})}\mathbf{x}}+C\mathrm{e}^{i\,\mathbf{k}_2^{(\mathrm{in})}\mathbf{x}}$, whereas the latter contributions to $(\frac{d\Omega}{d\sigma})_{\mathrm{X}}$ stem from the interference in the scattering of the incident double-beam. Note that the total scattering structure function on the right side of the proportionality relation (90) for $(\frac{d\Omega}{d\sigma})_{\mathrm{X}}$ is real.

We conclude by noting that the result (90) may be also written in the form [25]

$$\left(\frac{d\Omega}{d\sigma}\right)_{\mathrm{X}} \sim \left\langle \left| \sum_{i=1}^{N} \{\mathcal{F}(k_1)\mathrm{e}^{i\,\mathbf{k}_1\mathbf{r}_i} + C\mathcal{F}(k_2)\mathrm{e}^{i\,\mathbf{k}_2\mathbf{r}_i}\} \right|^2 \right\rangle , \tag{91}$$

where now as in (55)–(57) the thermodynamic expectation value $\langle\cdots\rangle = \int\cdots W(\mathbf{R},\mathbf{R};\beta)\,d\mathbf{R}$ in (91) is formed w.r.t. the canonical density operator (31) (for $\mathbf{R}'=\mathbf{R}$ therein) of the N atoms of the scattering target system, the positions of which in three-dimensional space $|\mathrm{R}^3$ are denoted by $\mathbf{r}_i$, $1\leq i\leq N$, in (91). The atomic form factors $\mathcal{F}(k_1)$ and $\mathcal{F}(k_2)$ in (91) with $k_1=|\mathbf{k}_1|$ and $k_2=|\mathbf{k}_2|$ are given by (81).

6 Summary and Future Prospects

In this work it was shown that the *full* two-body density $\varrho_2(\mathbf{S},\mathbf{r})$ of a crystalline material may be mapped out by means of diffuse scattering of a double-beam of neutrons or X-rays. The pertinent Bragg and diffuse scattering structure functions are defined in (48) and (49). These in turn are related to the differential cross sections in the static approximation for nearly elastic scattering of an incident double-beam of neutrons or X-rays via (26), (43), and (90). In order to obtain experimental information about the $\mathbf{K}\neq 0$ terms $u(\mathbf{r};\mathbf{K})$ in the lattice Fourier series (40) of the two-body density $\varrho_2(\mathbf{S},\mathbf{r})$ from the diffuse background scattering in two-beam scattering experiments, the difference $\mathbf{K}=\mathbf{k}_1^{(\mathrm{in})}-\mathbf{k}_2^{(\mathrm{in})}$ of the wave number vectors $\mathbf{k}_1^{(\mathrm{in})}$ and $\mathbf{k}_2^{(\mathrm{in})}$ in the incoming double-beam $\mathrm{e}^{i\,\mathbf{k}_1^{(\mathrm{in})}\mathbf{x}}+C\mathrm{e}^{i\,\mathbf{k}_2^{(\mathrm{in})}\mathbf{x}}$ must coincide with a reciprocal lattice vector $\mathbf{K}$ of the scattering crystal. We hasten to emphasize that such double-beam experiments can indeed be carried out [26–35], yielding experimental access to the *full* two-body density $\varrho_2(\mathbf{S},\mathbf{r})$.

Within the existing framework of exact group-theoretical FPIMC simulations of crystals [3–6] the Bragg scattering structure function $S_{\mathrm{B}}(\mathbf{K};\mathbf{k})$, defined in (48), may be readily obtained from the computed Fourier transform coefficients $\rho(\mathbf{K})$ of the local one-body density $\varrho(\mathbf{x})$. Thus the FPIMC approach of Refs. 3–6 allows for straighforward comparisons of theoretical results for $S_{\mathrm{B}}(\mathbf{K};\mathbf{k})$ with the dominant Bragg peaks in experimentally measurable differential cross sections of neutron scattering and X-ray diffraction from crystals. In this context we mention in particular the experimental single-crystal single-beam X-ray structure analysis of novel high-pressure high-temperature helium and xenon solids [36–38]. The Bragg peak intensities measured in these X-ray diffraction experiments are proportional to $|\rho(\mathbf{K})|^2$, which quantity may then be compared to numerical FPIMC results. Furthermore, in addition to the atomic one-body density $\varrho(\mathbf{x})$ the local one-electron density $\varrho_{\mathrm{e}}(\mathbf{x})$ of these inert gas crystals may be obtained from the computed $\rho(\mathbf{K})$ with the help of (79) and (80), resulting in $\varrho_{\mathrm{e}}(\mathbf{x}) = \sum_{\mathbf{K}} \rho(\mathbf{K})\mathcal{F}(K)\,\mathrm{e}^{i\,\mathbf{K}\mathbf{x}}$. Theoretical work in which the above mentioned quantities are computed in exact group-theoretical FPIMC calculations for high-pressure high-temperature helium solids is at present in progress.

The group-theoretical FPIMC approach [3–6] exploits the transformation properties (63), to cast the lattice Fourier transforms $u(\mathbf{r};\mathbf{K})$ of the two-body density $\varrho_2(\mathbf{S},\mathbf{r})$ in the form of expansions in terms of complete sets of Hilbert space basis functions of the relevant symmetries $\Gamma_{\tau,\mathcal{L}_0(\mathbf{K})}$, given by the mapping (62). This procedure [3–6] yields symmetry-adapted series representations of the $u(\mathbf{r};\mathbf{K})$ with expansion coefficient functions that depend on the relative distance $r = |\mathbf{r}|$ and on the polar angle ϑ that $\mathbf{r}$ encloses with the z-axis of the frame of reference. These expansions render the $u(\mathbf{r};\mathbf{K})$ in the shape of symmetrized trigonometric series in the azimuthal angle φ that $\mathbf{r}$ encloses with the x-axis with, however, $r\vartheta$-dependent expansion coefficient functions. It is these $r\vartheta$-dependent coefficient functions of the symmetrized series representations of the $u(\mathbf{r};\mathbf{K})$ which are computed numerically by means of the exact FPIMC integration techniques of Refs. 3 and 4. Therefore the diffuse scattering structure function $S_{\mathrm{d}}(\mathbf{K};\mathbf{k})$, defined in (49), may be calculated numerically by inserting in (49) the computed Fourier transform $\rho(\mathbf{K})$ of $\varrho(\mathbf{x})$ and the symmetrized trigonometric series representations of $u(\mathbf{r};\mathbf{K})$ with the computed $r\vartheta$-dependent coefficient functions. The Fourier transform $\mathcal{F}[u(\mathbf{r};\mathbf{K})](\mathbf{k})$ in (49) (cf. (46)) may be computed by standard numerical integration techniques. However, the FPIMC methods of Refs. 3 and 4 are also ideally suited to directly sample the total scattering operators $\sum_{i=1}^{N}\cdots$ in (55) and (91) and determine the Bragg and diffuse scattering structure functions by this method. Future group-theoretical FPIMC calculations along the lines sketched in this paragraph will be devoted to computing the Bragg and diffuse scattering structure functions for the scattering of double-beams of neutrons and X-rays for the novel high-temperature high-pressure helium and xenon solids investigated experimentally in Refs. 36–38.

Acknowledgments

This work was supported by United Kingdom EPSRC Grant GR/M45429: "Microscopic quantum many-body calculations of complex structures." The kind hospitality and financial support of the Graduiertenkolleg GRK 549 "Azentrische Kristalle" at the University of Cologne in Germany is appreciated. I am grateful for discussions and expert advice to L. Bohatý, H. Boysen, G. Eckold, F. Frey, D. Strauch, and in particular to W. Schülke and H. Zimmermann.

References

1. K. A. Gernoth, *in* "Proceedings of the 165. WE-Heraeus-Seminar, Theory of Spin Lattices and Lattice Gauge Models" (J. W. Clark and M. L. Ristig, Eds.), Lecture Notes in Physics, Vol. 494, p. 84, Springer-Verlag, Berlin, 1997.
2. K. A. Gernoth, *in* "Condensed Matter Theories" (J. W. Clark and P. V. Panat, Eds.), Vol. 12, p. 331, Nova Science, Commack, NY, 1997.
3. K. A. Gernoth, *Ann. Phys. (NY)* **285** (2000), 61.
4. K. A. Gernoth, *Ann. Phys. (NY)* **291** (2001), 202.
5. K. A. Gernoth, *in* "150 Years of Quantum Many-Body Theory" (R. F. Bishop, K. A. Gernoth, and N. R. Walet, Eds.), Series on Advances in Quantum-Many-Body Theory, Vol. 5., p. 249, World Scientific, Singapore, 2001.
6. K. A. Gernoth, *J. Low Temp. Phys.* **126** (2002), 725.
7. J. C. Tolédano and P. Tolédano, "The Landau Theory of Phase Transitions," World Scientific, Singapore, 1987.
8. L. D. Landau, "Collected Papers" (D. ter Haar, Ed.), 1st ed., p. 193, Pergamon, Oxford, 1965.
9. V. Apaja, J. Halinen, and M. Saarela, *J. Low Temp. Phys.* **113** (1998), 909.
10. J. Halinen, V. Apaja, K. A. Gernoth, and M. Saarela, *J. Low Temp. Phys.* **121** (2000), 531.
11. F. Pederiva, S. A. Vitiello, K. A. Gernoth, S. Fantoni, and L. Reatto, *Phys. Rev. B* **53** (1996), 15129.
12. K. A. Gernoth, *in* "Condensed Matter Theories" (D. J. Ernst, I. E. Perakis, and A. S. Umar, Eds.), Vol. 14, p. 461, Nova Science, Huntington, NY, 2000.
13. D. J. J. Farnell, R. F. Bishop, and K. A. Gernoth, *Phys. Rev. B* **63** (2001), 220402-1.
14. D. J. J. Farnell, K. A. Gernoth, and R. F. Bishop, *Phys. Rev. B* **64** (2001), 172409-1.
15. D. J. J. Farnell, R. F. Bishop, and K. A. Gernoth *J. Stat. Phys.* (2002), in press.
16. D. N. Timms, A. C. Evans, M. Boninsegni, D. M. Ceperley, J. Meyers, and R. O. Simmons, *J. Phys.: Condens. Matter* **8** (1996), 6665.
17. D. A. Peek and R. O. Simmons, *J. Chem. Phys.* **94** (1991), 3169.
18. D. A. Peek, M. C. Schmidt, I. Fujita, and R. O. Simmons, *Phys. Rev. B* **45** (1992), 9671.
19. D. A. Peek, M. C. Schmidt, I. Fujita, and R. O. Simmons, *Phys. Rev. B* **45** (1992), 9680.
20. S. W. Lovesey, "Theory of Neutron Scattering from Condensed Matter," Vol. 1, Clarendon Press, Oxford, 1984.
21. J. F. Cornwell, "Group Theory in Physics," Vol. I, 5th ed., Academic Press, London, 1994.
22. G. Burns and A. M. Glazer, "Space Groups for Solid State Scientists," 2nd ed., Academic Press, Boston, 1990.

23. "International Tables for Crystallography" (A. J. C. Wilson, Ed.), Vol. C, Kluwer Academic, Dordrecht, 1995.
24. "International Tables for Crystallography" (T. Hahn, Ed.), Vol. A, 4th Ed., Kluwer Academic, Dordrecht, 1996.
25. "International Tables for Crystallography" (U. Shmueli, Ed.), Vol. B, Kluwer Academic, Dordrecht, 1996.
26. W. Bauspiess, U. Bonse, W. Graeff, and H. Rauch, *J. Appl. Cryst.* **10** (1977), 338.
27. U. Bonse, *in* "Neutron Interferometry" (U. Bonse and H. Rauch, Eds.), p. 3, Clarendon Press, Oxford, 1979.
28. H. Rauch, *in* "Neutron Interferometry" (U. Bonse and H. Rauch, Eds.), p. 161, Clarendon Press, Oxford, 1979.
29. W. Schülke, U. Bonse, and S. Mourikis, *Phys. Rev. Lett.* **47** (1981), 1209.
30. W. Schülke, *Solid State Commun.* **43** (1982), 863; Erratum: *Solid State Commun.* **44** (1982), 1130.
31. G. Materlik, and J. Zegenhagen, *Phys. Lett. A* **104** (1984), 47.
32. H. Spalt, A. Zounek, B. N. Dev, and G. Materlik, *Phys. Rev. Lett.* **60** (1988), 1868.
33. U. Bonse, *Physica B* **151** (1988), 7.
34. W. Schülke and A. Kaprolat, *Phys. Rev. Lett.* **67** (1991), 879.
35. H. Rauch and S. A. Werner, "Neutron Interferometry: Lessons in Experimental Quantum Mechanics", Clarendon Press, Oxford, 2000.
36. A. P. Jephcoat, H. K. Mao, L. W. Finger, D. E. Cox, R. J. Hemley, and C. S. Zha, *Phys. Rev. Lett.* **59** (1987), 2670.
37. H. K. Mao, R. J. Hemley, Y. Wu, A. P. Jephcoat, L. W. Finger, C. S. Zha, and W. A. Bassett, *Phys. Rev. Let.* **60** (1988), 2649.
38. P. Loubeyre, R. LeToullec, J. P. Pinceaux, H. K. Mao, J. Hu, and R. J. Hemley, *Phys. Rev. Lett.* **71** (1993), 2272.

Average Structure vs. Real Structure: Molecular Dynamics Studies of Silica

Martin H. Müser

Institut für Physik, Johannes Gutenberg-Universität Mainz, Germany

Abstract. The microscopic structure of a crystal and thermal fluctuations of the atoms constituting the crystal are intimately connected with the macroscopic elastic properties including mechanical stability. In some cases, however, the picture is more complex than that which is drawn in text books on solid state physics. (i) The instantaneous microscopic structure can deviate in a non-Gaussian way from the average structure even when domain disorder and/or crystal defects are absent. Quasi harmonic approximations may then turn out to be meaningless. (ii) The crystal is subject to external pressures that are sufficiently large in order to render the definition of elastic constants non unique. These two points are discussed exemplarily in the context of the high-temperature and the high-pressure phases of quartz. In particular, it is discussed how to observe and how to classify non-Gaussian disorder in molecular dynamics (MD) simulations and how to evaluate mechanical stability of solids under pressure. Some details are given on the calculation of thermal, mechanical, and structural properties of solids, also for temperatures far below their Debye temperature.

1 Introduction

Pure silica (SiO_2) has a rich phase diagram with interesting temperature and pressure induced phase transitions between the various stable or metastable polymorphs[Dolino 1990], [Heaney, Prewitt, and Gibbs 1994]. The large variety of crystalline phases is due to the tetrahedral structure of relatively rigid SiO_4 units and the large geometric flexibility with which they can be connected. The material properties in general and the mechanisms driving the phase transition in particular are often explained in terms of their average structure and (Gaussian) fluctuations around this average structure. Detailed molecular dynamics simulations reveal that the picture needs to be refined [Müser and Binder 2001]: In the high-temperature phases, e.g., in β-quartz, the fast oscillations of oxygen atoms are around (time-dependent) positions that do not correspond to the ideal oxygen positions in β quartz. The averaged configurations only resemble the ideal structure if averaged over time scales that are distinctly larger than typical inverse phonon frequencies. This effect has serious implications on the calculation of material properties: The calculation of elastic constants as evaluated in terms of a quasi-harmonic approximation turns out to be meaningless.

Another controversially discussed phase transition of quartz is the pressure-induced transformation of α-quartz into quartz-II [Kingma et al. 1993]. Based on a mechanical stability analysis by means of molecular dynamics (MD) simulations, Binggeli and Chelikowsky (1992) suggested that the transformation is induced by an elastic instability. The MD predictions were in contradiction with an experimental mechanical stability analysis by Gregoryanz et al. (2000). The discrepancy between simulation and experiment may not be due to poor MD modeling but it may rather be due to the ambiguity in defining elastic constants for solids under large pressure [Müser and Schöffel 2001]. Of course, the correct stability criterion must require that the long-range structural fluctuations are finite.

In these lecture notes, a selective review on MD simulations of various quartz modifications and the phase transformations between them is given. In particular, it is emphasized what a computer simulation should provide in order to be compared to experiments and in order to make predictions that complement reliably experimental observations. The potential energy surfaces suggested by van Beest, Kramer, and van Santen (1990) is used as their two-body potential turned out to describe rather accurately various aspects of silica in different crystalline phases [Tse and Klug 1991].

2 Molecular Dynamics Simulations of Crystals

The code for a molecular dynamics (MD) simulation [Allen and Tildesley 1987], [Frenkel and Smit 1996] or a Monte Carlo (MC) [Landau and Binder 2000] simulation of a crystal is of course similar to that of any system which is described in terms of effective or *ab initio*-based interatomic potentials. In principle, the starting point is a Lagrangian L

$$L = \sum_{n=1}^{N} \frac{1}{2} m_n \dot{\mathbf{R}}_n^2 - V(\{R\}), \tag{1}$$

where m_n denotes the mass of particle n and $V(\{R\})$ is the net potential energy as a function of the coordinates of all N particles. The well-known Newton equation's of motion for the particles are derived from L. These equations are integrated numerically, stepping forward in time by discrete steps of size Δt. To thermostat the system, the equations of motion are modified so that the average kinetic energy stays at its equilibrium value. There are various approaches to achieve this. An efficient way that also works without modifications for strongly harmonic systems (note that the famous Nosè-Hoover thermostat suffers from serious difficulties for such systems!) is to couple each atom to its own local thermostat [Schneider and Stoll 1978]. The exchange of energy with the outside world is modeled by a Langevin equation that includes a damping coefficient γ and a random force $\mathbf{\Gamma}_n(t)$ on each atom. The equations of motion for the α component of the position $R_{n\alpha}$ become:

$$m_n \frac{d^2 R_{n\alpha}}{dt^2} = -\frac{\partial}{\partial R_{n\alpha}} V(\{R\}) - m_n \gamma \frac{dR_{n\alpha}}{dt} + \Gamma_{n\alpha}(t). \tag{2}$$

In order to satisfy the fluctuation-dissipation theorem, the $\Gamma_{n\alpha}(t)$ must be completely random, have zero mean, and have a second moment given by

$$\langle \Gamma_{n\alpha}(t)\Gamma_{n'\beta}(t')\rangle = 2k_{\mathrm{B}}T\gamma m_n \delta_{nn'}\delta_{\alpha\beta}\frac{\delta_{tt'}}{\Delta t}. \qquad (3)$$

Note that (3) is written down for a discretized time.

While the remarks above are generally valid, there are two important points that are usually more relevant for crystals than for fluids or other disordered systems. (i) The crystal structure may not be conform with a periodically repeated simulation box of orthorhombic symmetry. This makes it necessary to employ arbitrary parallelepiped shaped simulation boxes. Such geometries also enable us to calculate all elastic constants from thermal strain fluctuations. (ii) Quantum effects typically play a more significant role in crystals than in fluids, although fluid helium is a prominent exception of this rule. Ways to incorporate these two points into MD simulations will now be discussed in further detail.

2.1 MD for Arbitrary Parallelepiped Simulation Cells

The simulation of crystals in an arbitrarily parallelepiped shaped simulation cell goes back to an idea invoked by a series of papers by Parrinello and Rahman (1980, 1981, and 1982): The atomic positions are represented as a scalar product of a dimensionless vector $\mathbf{r}$ (with components between zero and unity) with a (symmetric!) matrix h whose rows are parallel to the three edges spanning the simulation cell:

$$R_{n,\alpha} = \sum_{\beta=1}^{3} h_{\alpha\beta}\, r_{n,\beta}. \qquad (4)$$

The Lagrangian described in (1) is then generalized by also attributing inertia to the simulation cell's geometry and by coupling the simulation cell's volume $\det(\mathsf{h})$ to an isotropic pressure p:

$$L\left(\mathsf{h}, \dot{\mathsf{h}}, \{\mathsf{r}\}, \{\dot{\mathsf{r}}\}\right) = \sum_{\alpha\beta} \frac{1}{2} W \dot{h}_{\alpha\beta}^2 + \sum_{n=1}^{N} \frac{1}{2} m_n (\mathsf{h}\dot{\mathsf{r}}_n)^2 \; - V(\{\mathsf{h}r\}) - p\det(\mathsf{h}). \quad (5)$$

Note that the original Lagrangian is recuperated by setting $\dot{\mathsf{h}}$ to zero and that the choice of the inertia term W is not unique, because this term can not be determined from first principles. While the choice of W affects the dynamics of the system, it does not affect the distribution functions of h and $\{\mathbf{r}\}$.

Using the Lagrangian formalism, the equations of motion for h and $\{\mathbf{r}\}$ can be derived in a straightforward way. It is convenient to thermostate all variables in order to have quickly converging ensemble averages. The choice of a Langevin type thermostat for h is particularly helpful as these modes are usually strongly harmonic and hence couple only weakly to the inner degrees of freedom $\{\mathbf{r}\}$.

In order to define the strain tensor ϵ, we need to know the thermal expectation value $\mathsf{h}_0 = \langle \mathsf{h} \rangle$. ϵ is given by

$$\epsilon = \frac{1}{2}\left(\mathsf{h}_0'^{-1}\mathsf{h}'\mathsf{h}\mathsf{h}_0^{-1} - 1\right). \qquad (6)$$

Thus by monitoring the various moments of h in the simulation, it is possible to calculate the first and the second moment of the thermal strain fluctuations $\delta\epsilon$ from which the (isothermal) elastic constants $C^T_{\alpha\beta\gamma\delta}$ follow via:

$$(C^T)^{-1}_{\alpha\beta\gamma\delta} = \frac{\det(\mathsf{h}_0)}{k_B T} \langle \delta\varepsilon_{\alpha\beta}\delta\varepsilon_{\gamma\delta}\rangle . \tag{7}$$

If this fluctuation relation is applied to non-zero pressures, then the $C_{\alpha\beta\gamma\delta}$ correspond to Birch coefficients rather than to elastic constants. This point will be further elucidated in Sect. 5. In any case, (7) is a manifestation of the connection between structure fluctuations and elastic properties.

2.2 MD Below the Debye Temperature

Path integral Monte Carlo (PIMC) [Barker 1979] and path integral molecular dynamics (PIMD) [Tuckerman et al. 1993] have proven useful in the atomistic simulation of quantum effects occurring in condensed matter at low temperatures (see also the article by Gernoth in these Lecture Notes). The application of path integral simulations is not restricted anymore to the calculation of thermal and structural properties of Lennard-Jones type systems but the treatment of more complex condensed matter systems becomes increasingly feasible.

In order to derive the path-integral scheme for a system with fluctuating simulation cell size, it is convenient to start from the Lagrangian given in (5). One should not quantize the tensor h, because in the long-wavelength limit this quantity becomes always classical and because the choice of the inertia W is arbitrary. Thus the quantization of the variables $\mathbf{r}_n$ can be achieved as usual. Path integral simulations exploit Feynman's idea to represent the partition function of a quantum mechanical point particle $Z(\beta)$ as a partition function of a classical ring polymer [Feynman and Hibbs 1965]. The position of a quantum mechanical point particle is represented by a chain with coordinates $\mathbf{r}_t$ with $t = 1, ..., P$ and cyclic boundary conditions $\mathbf{r}_t = \mathbf{r}_{t+P}$. In the present notation, the partition function reads:

$$Z(\beta) = \lim_{P\to\infty} \frac{[\det(\mathsf{h})]^{NP}}{\prod_{n=1}^{N} \lambda_n^{3P}(\beta/P)} \prod_{n=1}^{N}\prod_{t=1}^{P} \int_0^1 d^3 r_{n,t} e^{-\beta V_{\rm rp}/P} \tag{8}$$

with

$$V_{\rm rp}(\{R\}) = \sum_{t=1}^{P} \left[V(\mathbf{R}_{1,t}, ..., \mathbf{R}_{N,t}) + \frac{1}{2}\sum_{n=1}^{N} \frac{m_n P^2}{\beta^2\hbar^2} \left(\mathbf{R}_{n,t} - \mathbf{R}_{n,t+1}\right)^2 \right] , \tag{9}$$

where the relation between $\mathbf{r}_n$ and large $\mathbf{R}_n$ is given as usual by (4). V represents the (real) potential energy of the system evaluated for particle coordinates at "imaginary time" t and the last summand of the right-hand side of (9) reflects the harmonic springs that keeps monomer of the ring polymer close together. It is instructive to visualize the interactions described in (9), which is done in Fig. 1.

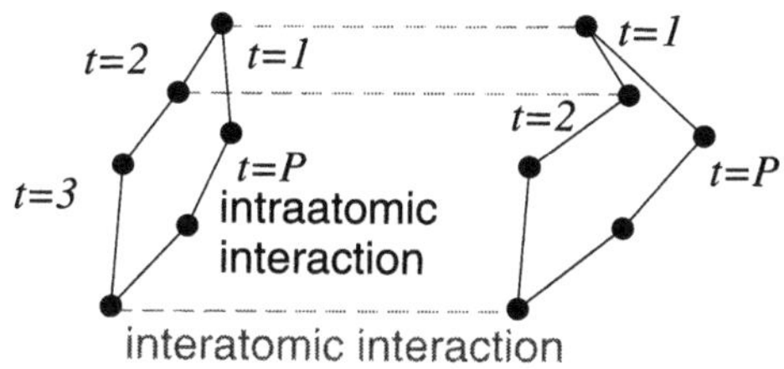

Fig. 1. Illustration of the interaction between two quantum mechanical point particles that are represented as classical ring polymers

The spring constants k_n connecting neighboring beads within a chain are given by $k_n = m_n P^2/\beta^2\hbar^2$. Hence large temperatures and/or large masses lead to a localization of the ring particle and thus to classical behaviour. The spatial extension of the chains at low temperatures reflects the thermal de Broglie wavelength.

In the quantum limit $P \to \infty$, the k_n become very stiff. This would lead to technical difficulties in a PIMD simulation if the "kinetic" masses of the beads were all chosen to be identical, because there would be a time scale separation between the center-of-mass mode and the internal modes. This effect would automatically result in an inefficient sampling, which is why one needs to come up with more efficient algorithm [Tuckerman et al. 1993]. One possibility is to attribute the inert masses (note that the real, physically meaningful masses are reflected in the harmonic springs) to the eigenmodes of the free chain and to chose them such that internal eigenfrequencies and typical frequencies associated with the center-of-mass motions are similar [Müser 2001]. For a proper choice of these masses including W and a proper choice for the Langevin damping coefficient γ, it is possible to obtain quickly converging estimates of many thermal and structural properties including the elastic constants.

3 Quantum Effects in α-Quartz

Many potential energy surfaces are adjusted such that they yield the proper lattice constants (parameters) and the correct elastic properties at low temperatures. This is also the case for the so-called BKS potential suggested for SiO_2 by van Beest, Kramer, and van Santen (1990). It is therefore an important test for a model potential surface to yield the correct thermal expansion at low temperature, because the anharmonic interactions are not explicitly incorporated into the potential parameters. Such a test requires a quantum mechanical treatment of the ionic motion, since classical expansion coefficients α remain finite as the temperature T tends to zero, while for a quantum mechanical treatment – like in experiment – α vanishes as T approaches zero. Despite the existence of sophisticated quasi-harmonic theories, it remains a challenge to predict reliably lattice parameters near absolute zero. Path integral simulations like PIMD, however, achieve very good resolution down to temperatures well below the Debye temperature T_D without any uncontrolled approximations other than the

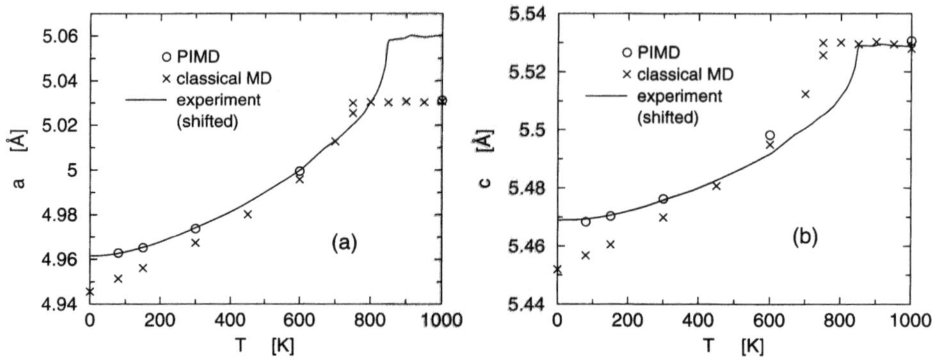

Fig. 2. Lattice constants of α-quartz at ambient pressure as a function of temperature. a) Experimental values for the a-axis are shifted to larger values by 0.06 $\AA$. b) Experimental values for the c-axis are shifted to larger values by 0.07 $\AA$. Error bars of simulations in all cases smaller than 100 fm. Experiment taken from Carpenter et al. (1998). From Müser (2001)

uncertainties due to the potential energy surface. Of course, these uncertainties would also be present in theoretical treatments.

Quartz has two independent lattice constants, a and c, with a and c representing the lattice parameters parallel to the x and z-direction respectively (in the standard representation). It turns out that thermal expansion is very well reproduced by the BKS potential at very low temperatures as can be seen in Fig. 2, while classical simulations show the wrong trend at small T.

Some numerical approaches (like density functional theory or other quantum chemistry approaches) reach a nearly perfect agreement with experiment, especially for structural properties. In general, such calculations do not include quantum effects of the ionic motion. As we just saw, quantum effects lead to an equilibrium structure that slightly differs from the "classical" equilibrium structure. From Fig. 2 we can learn that a highly accurate calculation should actually underestimate the experimentally measured lattice parameters for quartz near absolute zero by about 0.35 %. Of course, finite-temperature simulations also provide important tests, i.e., one observes a discrepancy in the α-β phase transition temperature of nearly 100 K and more importantly, the jump in the lattice parameter a is strongly suppressed in the simulation. These shortcomings are discussed elsewhere [Müser and Binder 2001].

A further phenomenon which needs quantum effects of the ionic motion to be considered are the well known isotope effects. E.g., for rare gas crystals such as ^{20}Ne and ^{22}Ne systematic differences between the lattice parameters were measured. The differences were rather nicely reproduced by path integral Monte Carlo simulations [Müser et al. 1995]. However, isotope effects will not be discussed here further.

The analysis of the influence of quantum mechanical effects is not limited to structural or thermal properties. With current computers and algorithms, it

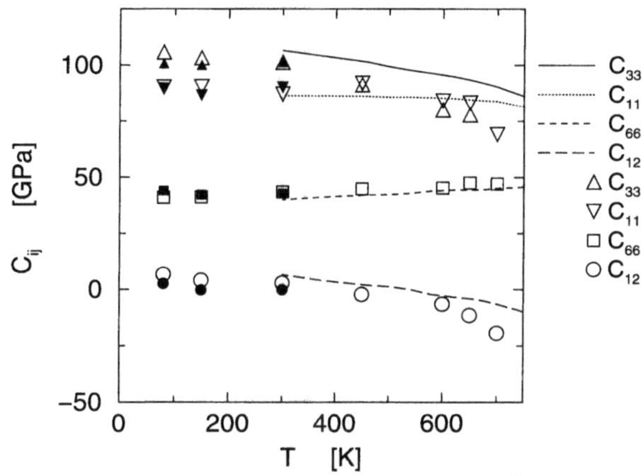

Fig. 3. Various elastic constants. Experimental data shown as curves is taken from Carpenter et al. (1998). Open symbols refer to classical simulations, filled symbols to PIMD simulations. Statistical error bars are about 2 GPa. From Müser (2001)

has also become possible to estimate elastic constants quite accurately, see i.e. the treatment of solid ^{3}He by Schöffel and Müser (2001). It seems a general trend that the *relative* corrections due to the quantum mechanical nature of ionic motion to cohesion energies (near $T = 0$ K) is larger than those to the lattice parameters and that the relative corrections to elastic constants is again larger than but in the same order as those to the cohesion energy. In α-quartz, the quantum induced reduction in the elastic constant C_{33} was estimated to be close to 5 GPa, which seems surprisingly large given the strong ionic-covalent bond in quartz and the relatively large masses of the constituting atoms. Figure 3 shows the elastic constants of α-quartz for a classical treatment, the quantum mechanical treatment along with some available experimental data.

Although this is hardly recognizable from the figure (because of the large scale needed for the ordinate), the temperature dependence of elastic constants calculated classically is qualitatively wrong: classical statistical mechanics always yields $\partial C_{ij}/\partial T \to$ const. as $T \to 0$ and this constant will in general be nonzero. Quantum mechanics requires $\partial C_{ij}/\partial T \to 0$ as $T \to 0$ due to the third law of thermodynamics.

From Fig. 3 it is noticeable that the elastic constants measured experimentally at 300 K match the elastic constants from the classical simulations at zero temperature.

This is due to the fact that the BKS potential was constructed such that ab-initio calculations were combined with bulk properties in order to fit the free model parameters. In the latter part, lattice constants and elastic constants were calculated for a classical system at $T = 0$ K from the (fit) parameters and adjusted such that agreement with experimental "quantum mechanical" (finite temperatures) data was optimum. While I consider the BKS potential to be the

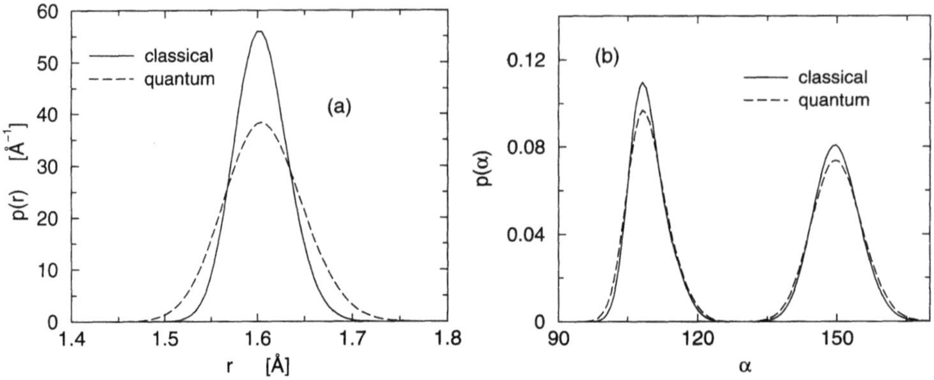

Fig. 4. a) Probability density $p(r)$ to find an oxygen atom in a distance r from a silicon atom. b) O-Si-O (left) and Si-O-Si (right) bond angle distribution function $p(\alpha)$. Solid lines reflect classical simulations, dashed lines represent quantum mechanical simulations. Temperature $T = 300$ K. From Müser (2001)

best two-body potential available for the simulation of bulk SiO_2, one can see that there is still some room left for improvement of the potential parameters.

It is interesting to note that T_D of α-quartz as determined by specific heat measurements [Striefler and Barsch 1975] is a strongly temperature-dependent function: At $T = 0$, $T_D \approx 550$ K, while at room temperature $T_D \approx 1,000$ K. This unusual behaviour can be understood if one keeps in mind the relevant degrees of freedom: There are low-lying excitations associated with so-called rigid unit modes (RUM) [Axe and Shirane 1970]. The RUM's are (collective) motions of stiff tetrahedral SiO_4 units invoking bending of the SiOSi bonds. As a matter of fact, it is possible to find support for this picture by simply comparing classical and quantum mechanical distribution functions of the SiO bond length and the SiOSi and OSiO bond angle distribution at room temperature, which is done in Fig. 4. The quantum mechanical Si-O bond length deviates considerably from the classical bond length distribution, while the bending of both SiOSi and OSiO angle is still classical, i.e. low-energetic. The bending modes start freezing quantum mechanically at temperatures near 200 K, while the SiO bond length distribution remains essentially unaltered upon further cooling. Particularly strong quantum effects in distributions $p(r)$ or $p(\alpha)$ such as shown in Fig. 4 are found in cases where the crystal contains very light atoms, e.g. orthorhombic polyethylene C_nH_{2n+2} [Martonak et al. 1998].

4 The Structure of β-Quartz

The average structure of β quartz is hexagonal and the symmetry class is P622. However, the local structure is known to deviate considerably from the average structure and the nature of this deviation has been a subject of long debate. It was frequently discussed in reference to the nature of the α-β transition in quartz [Dolino 1990,Heaney, Prewitt, and Gibbs 1994,Carpenter et al. 1998]. The main issue is whether or not oxygen atoms oscillate around positions that correspond to locations on which oxygen atoms can be found in α quartz, so-called α_1 and α_2 positions. If β quartz consisted of small α_1 and α_2 domains (each occuring with the same probability), then the averaged structure would still be that of β quartz. Alternatively, one may envision a situation where the atoms oscillate around their β positions in such a way that the short-range order differs from that in α-quartz. Evidence for this was found by Kihara (1990), who suggested vibrational motion of the oxygen atoms around the Si-Si lines leading to non Gaussian atomic distribution functions.

In the following discussion, I will mostly disregard the intermediate incommensurate phase, which has been observed in a relatively small temperature range of 1.5 K [Dolino 1990] between the regions where α and β quartz are thermodynamically stable, respectively. The wavelength of the incommensurate soft mode exceeds the feasible linear dimension of our current atomistic simulation by more than an order of magnitude. Hence if we were to maintain the shape of our simulation box, we would have to deal with system sizes more than three orders of magnitude larger than those that are currently employed. Going to such large system sizes would be barely feasible at the present time on a cluster containing a few workstations even with further simplifications of the model and optimization of the algorithm (replace Ewald sums with algorithms that are preferential for summing up Coulombic interactions of extremely large systems). One might argue that it could be sufficient to extend the size of the simulation cell parallel to the soft mode vector. However, this could change the effective dimensionality of our solid, which is likely to invoke a qualitatively different phase transition scenario. Note that studying relatively small systems of a few thousands of atoms is not necessarily a disadvantage. It has been argued that the α-β transition is first order due to the existence of the incommensurate phase in between the two phases. Our system sizes are too small to be affected by this intermediate phase, however, they are still large enough to determine the order of the transition via finite-size scaling.

I will first be concerned with a discussion of how to quantify and to observe the non-Gaussian behaviour of β quartz in atomistic computer simulations. One may expect that the behaviour is rather generic for network formers with tetrahedral short-range order and displacive phase transition between high symmetry (i.e. β-quartz) and low symmetry (i.e. α-quartz) phases. We will then explore the consequences that non-Gaussian behaviour has for quasi harmonic treatments. This will be done in Sect. 4.3. The relation of the local order in β-quartz with respect to that in α-quartz will be discussed in detail in Sect. 4.3.

4.1 Non-gaussian Disorder in β-Quartz

Let us first discuss the information that is accessible to an experimentalist and explore possibilities how computer simulations can retrieve similar data or even go beyond. The key quantity in structure determination by elastic scattering (neutron or X-ray) is the scattering intensity $I(\mathbf{q})$

$$I(\mathbf{q}) = \sum_{ij} b_i b_j \langle \exp\{i\mathbf{q}(\mathbf{R}_i - \mathbf{R}_j)\}\rangle, \tag{10}$$

with $\mathbf{q}$ the momentum transfer of the scattered neutron, b_i the cross section of atom i, and $\mathbf{R}_i$ its position in space. $\langle ... \rangle$ denotes a temporal average over a sample in thermal equilibrium. If the coupling between different vibrational modes in the crystal is weak, (10) can be simplified to

$$I(\mathbf{q}) = I_{\text{ideal}}(\mathbf{q})\tilde{S}(\mathbf{q}), \tag{11}$$

where $I_{\text{ideal}}(\mathbf{q})$ is the scattering intensity for the ideal structure (or average structure) and $\tilde{S}(\mathbf{q})$ is a weight factor, also known as Debye Waller factor, that depends on the amount of vibrational or structural disorder in the system. Unlike experiment, computer simulations can determine $I_{\text{ideal}}(\mathbf{q})$ by first averaging the configurations over many time steps and then performing the scattering, i.e., the brackets $\langle ... \rangle$ are pulled into the argument of the exponential. Of course this procedure is limited to time scales in which atomic diffusion within the solid can be neglected. Since in most solids the diffusive mechanisms take place on time scales distinctly larger than those accessible in molecular dynamics simulations, this effect can usually be neglected. Exceptions to this rule are quantum solids like crystalline helium and crystals near a (non-displacive) phase transition. In both cases significant diffusion can take place on the time scale of a MD simulation.

The dependence of $\tilde{S}(\mathbf{q})$ on q (the absolute value of $\mathbf{q}$) can be easily discussed in a one-dimensional system. This is done in Fig. 5 where the effect of phonons (Gaussian atomic distribution functions around the ideal positions) and disorder on the $\tilde{S}(\mathbf{q})$ is analyzed. The reader might consult standard solid state physics books in order to confirm the validity of the equations stated in that figure or derive them by using (10) and by decomposing the vibrations into harmonic eigenmodes. The main result for vibrational disorder is that the intensity of $I(q)$ is the product of the ideal scattering intensity $I_{\text{ideal}}(q)$ times a weight function that falls off exponentially fast with the square of the scattered wave vector $\mathbf{q}$, while oscillations in $I(\mathbf{q})$ are indicative of structural disorder. Of course, the last row in Fig. 5 is highly idealized, because the atoms will oscillate around the split positions and because real crystals are not one-dimensional.

The qualitative discussion in Fig. 5 is also valid for solids with basis, i.e. two-component systems like quartz. In order to generalize the analysis of how $\tilde{I}(\mathbf{q})$ depends on the inverse neutron wavelength $\mathbf{q}$, it is convenient to introduce the polarization vector $\mathbf{u}_{m\alpha}(\mathbf{Q})$ of the normal mode $(\mathbf{Q}, \alpha)$. The number of independent polarization vectors in a unit cell depends on the crystal symmetry.

In a computer simulation, we can restrict the evaluation of the scattering intensities, see (10), to pairs of atoms, which all have the same polarization vector $\mathbf{u}_{m\alpha}(\mathbf{Q})$ and $\mathbf{u}_{n\alpha}(\mathbf{Q})$, respectively. This enables us to define a generalized

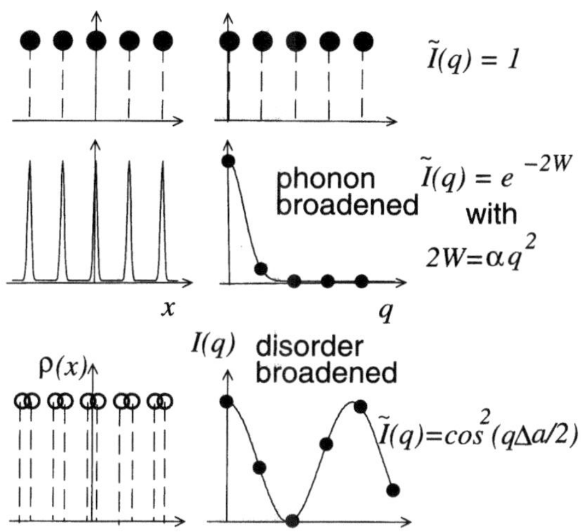

Fig. 5. Schematic representation of the density distribution $\rho(x)$ in real space (left) and elastic scattering intensity $\tilde{I}(q)$ (right) for a one-dimensional one-component system. The top row represents the ideal/average structure, the middle row a phonon broadened structure, and the bottom row shows a disordered structure in which one (split) atom occupies randomly one out of two equivalent positions that are separated by Δa.

scattering intensities or generalized Debye-Waller factor $I_{mn}(\mathbf{q})$:

$$I_{mn}(\mathbf{q}) = \exp\left\{ -q^2 \sum_{\mathbf{Q}\alpha} \langle [\mathbf{n_q} \mathbf{u}_{m\alpha}(\mathbf{Q})] \, [\mathbf{n_q} \mathbf{u}^*_{n\alpha}(\mathbf{Q})] \rangle \right\} . \tag{12}$$

While this generalized Debye-Waller factor is directly accessible in simulations, experimental determination of that quantity requires fitting and *a priori* assumptions on the atomic distributions. The $I_{mn}(\mathbf{q})$ have the important feature that they fall off according to $I \propto e^{-\alpha q^2}$ as long as the harmonic approximation is valid. The prefactor α in the exponential depends of course on the director $\mathbf{n_q}$ of $\mathbf{q}$. The property $I \propto e^{\alpha q^2}$ is lost if the sum in (10) is evaluated over all atoms. The following example, see Fig. 6, shows how the calculation of I_{mm}'s can contribute to the determination of the local structure in β quartz.

In Fig. 6, the (generalized) scattering intensities of the ideal structure (i.e. the structure averaged over a few 10,000 MD steps), the averaged scattering intensities I_{mm}, and fits are shown. The fits include an adjustable offset that is added to the Gaussian, i.e. $\tilde{I}_{\text{fit}} = e^{-\alpha q^2} + \text{const}$. Only those $\mathbf{q}$ vectors are incorporated in the fit that correspond to "allowed" lattice vectors. The offset was observed to vanish with inverse particle number. The background at the symmetry forbidden q can be understood from the fact that the simulations were done in the isothermal-isobaric ensemble. One can see in Fig. 6 that the generalized scattering intensities $I_{mm}(\mathrm{q})$ are quite well described by the harmonic approximation if m denotes one of the two equivalent Si atoms. (The Si atoms are equivalent in the sense that they have the same polarization vector for an eigenmode $\mathbf{Q}, \alpha$.)

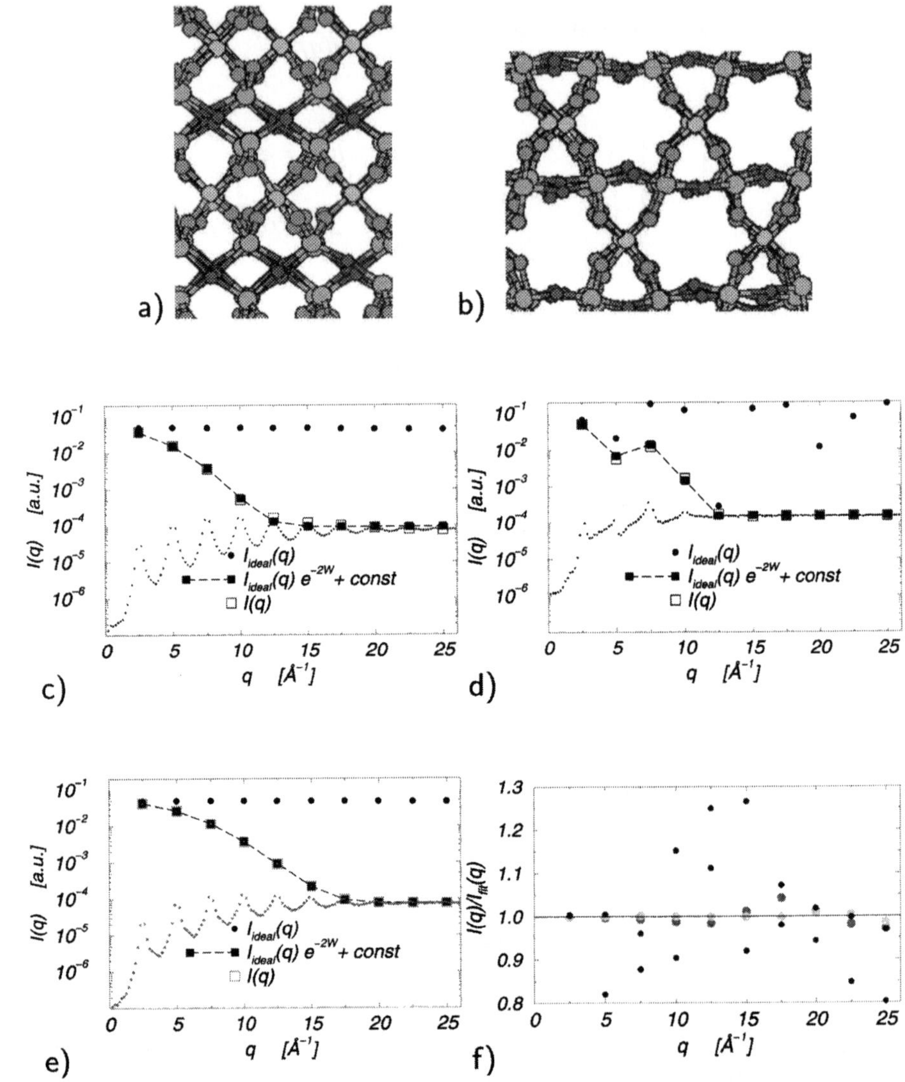

Fig. 6. a) Snapshot of β quartz at $T = 1000$ K in the xz plane. x is horizontal. Ocher and grey colored atoms reflect Si atoms, red and blue show O atoms. b) Same as a) but for xy plane. c)-e) Generalized scattering intensities (ideal structure, real structure, and fits) as a function of q with $\mathbf{n}_q$ parallel to x axis. Colors are consistent with a) and b). The background in $I(q)$ [small circles in c)-e)] including the plateau value for large q becomes smaller with increasing system size. f) Observed scattering intensities $I(q)$ divided by fit values $I_{\rm fit}(q)$

The oxygen atoms show much larger deviations from harmonic behaviour than the silicon atoms, since the oxygen's $I_{mm}(q)$ can be fit considerably less well with Gaussians.

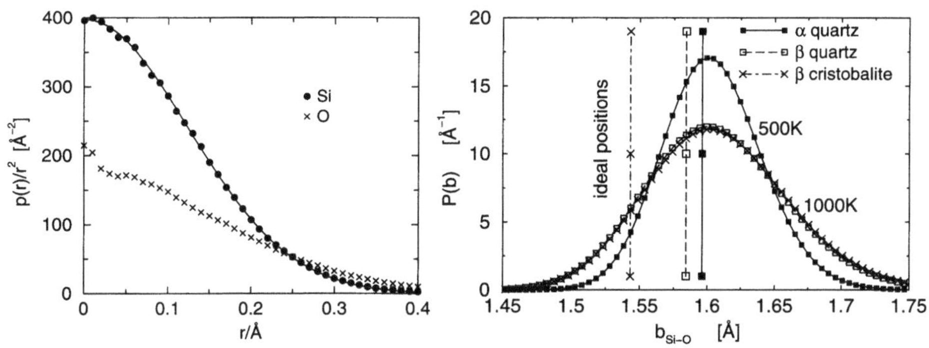

Fig. 7. Left: Probability $p(r)$ to find an atom a distance r away from its average position (with respect to the center of mass of the simulation box) divided by r^2 as a function of r. Circles refer to Si atoms, crosses to O atoms. The straight line is a Gaussian fit through the Si data. $T = 900$ K. Right: Probabilty $p(b)$ to find an oxygen atom a distance b away from a silicon atom. β-quartz and β-cristobalite simulations were carried out at $T = 1000$ K, those for α-quartz at 500 K. The straight lines reflect the location of the Si-O bond lengths as deduced from the average structure. From Müser and Binder (2001)

The non-Gaussian nature of the vibrations associated with the motion of oxygen atoms is also illustrated in Fig. 7, where the probability $p(r)$ to find an atom a distance r away from its average position (with respect to the center of mass of the simulation box) divided by r^2 is shown as a function of r. For the Si atoms, a single Gaussian is obtained, while the O atoms apparently have several preferred sites, which is obvious from the anomaly in the $p(r)/r^2$ curve. These results strongly support the X-ray study by Kihara (1990) in which the oxygen distribution functions were conjectured to deviate considerably from Gaussians.

Yet another indication for the fluctuations about the average positions being anharmonic is obtained from the bond-length distribution function. For β-cristobalite, it has been shown experimentally that the real bond-length distribution function peaks at a radius that is distinctly larger than the bond-length deduced from the average structure [Dove et al. 1997]. This effect is less strong in β-quartz as shown in Fig. 7. Nevertheless it is clear that the tendencies in β-cristobalite and β-quartz are similar: The Si-O bond lengths deduced from the average structures is located at a position that is markedly smaller than the position where the bond length distribution peaks. The bond-lengths from the average positions shown in Fig. 7 are deduced from our simulations (see also Fig. 8). The values we obtain for β-cristobalite agree well with those suggested by Dove et al. (1997). They state that the bond length of the average position is about 1.55 Å, while their bond length distribution peaks at 1.61-1.62 Å. There is also qualitative agreement of our simulations with experimental data on β-quartz: Kihara (1990) reported a real SiO bond length of 1.62 Å, which is about 0.04 Å larger than the spectroscopic bond length of 1.588 Å, while our

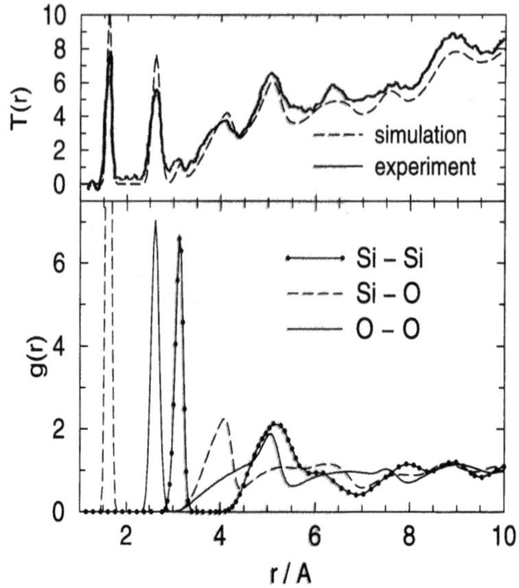

Fig. 8. a) $T(r)$ for β-cristobalite as a function of distance r at 573 K. Experiment is represented by a solid line and simulation data is represented by a dashed line. b) Corresponding radial distribution function $g(r)$ for Si-Si, Si-O, and O-O bonds. The curves are normalized such that $g(r) \to 1$ for $r \to \infty$. From Rickwardt et al. (2001)

simulations suggest a change of only 0.02 Å. A recent neutron diffraction study confirms Kihara's results quite accurately [Tucker et al. 2000].

While we are mainly concerned with β-quartz in this chapter, it should not remain unmentioned that simulations based on the BKS potential yield excellent agreement for the instantaneous order in other polymorphs as well, i.e. β-cristobalite. It is a cubic SiO_2 polymorph in which the Si are four-fold coordinated like in quartz.

Information on the instantaneous order can be obtained experimentally by measuring the so-called total pair correlation function $T(r)$. This function is a sum of SiSi, SiO, and OO pair correlation functions $g(r)$, whereby each function is weighted with the cross section b of the atoms involved, hence for SiO_2:

$$T(r) \propto b_{\mathrm{Si}} b_{\mathrm{Si}} g_{\mathrm{SiSi}}(r) + 2 b_{\mathrm{Si}} b_{\mathrm{O}} g_{\mathrm{SiO}}(r) + 4 b_{\mathrm{O}} b_{\mathrm{O}} g_{\mathrm{SiO}}(r). \tag{13}$$

I refer to Ref. [Dove et al. 1997] for technical details on how to obtain $T(r)$ experimentally. Again, the simulations provide a unique way to determine the individual $g(r)$'s directly whithout any assumptions on the structure or any kind of *a priori* assumptions. The results for $T(r)$ as obtained for β-cristobalite are shown in Fig. (8a) while the radial distribution functions are displayed in Fig. (8b). It is interesting to note that the peak in $T(r)$ at about 5 $\AA$ is due to a simultaneous maximum in $g_{\mathrm{O\,O}}(r)$ and $g_{\mathrm{Si\,O}}(r)$, while the peak in $T(r)$ at 6.25 $\AA$ does not have a corresponding peak in any $g(r)$. The large local maximum in $T(r)$ at about 9 $\AA$ is then located at a position where all $g(r)$'s have a local maximum as well. The shape of $T(r)$ is particularly sensitive to details of the potentials at large distances, e.g., cutting off the short-range part of the potential at 4.5 $\AA$ alters $T(r)$ significantly for $r \geq 8$ $\AA$.

4.2 Implications for Quasi-harmonic Treatments

Here, I want to argue that the seemingly small deviation from harmonic behaviour – as seen in the previous subsection – has dramatical consequences for quasi-harmonic treatments of the high-temperature phase of quartz. Of course, one may also expect similar difficulties with other high-temperature phases of network formers with similar tetrahedral short-range order. Quasi-harmonic approximations are done by expanding the potential energy surface about the experimentally observed ideal reference structure. From the bond length distribution one can see that this reference structure is not the most likely structure and hence the outlined expansion risks to fail in providing reliable structural and elastical properties.

Table 1 contains the list of various structural and elastic properties of α-quartz and β-quartz at temperatures $T = 300$ K and $T = 1000$ K, respectively. I compare experiments with molecular dynamics simulations carried out at finite temperatures and include data as obtained from harmonic approximations. The results for the harmonic approximations contain a treatment in which the BKS potential is used and lattice constants and elastic moduli are evaluated under the constraints that the silicon atoms remain on ideal β-quartz lattice position. The oxygen atoms were allowed to move freely, i.e. to relax to their ideal lattice positions and fluctuate around them. This procedure is strongly related to the way in which the elastic constants (and lattice parameters) are obtained in a local density approximation (LDA) calculation by Demuth et al. (1999) whose results are inserted as well for comparison.

It can be seen that structural parameters and elastic constants are in good agreement for the low-temperature phase in all cases. This speaks in favor for both the BKS potential and the LDA calculation by Demuth et al. In the high-temperature phase, however, only the molecular dynamics simulations that implicitly include the non-Gaussian lattice deformations agree satisfactorily with experiment while both harmonic treatments predict a bulk modulus which is about a factor two too large. Unfortunately no data was provided by Demuth et al. (1999) on other elastic properties. The disagreement between harmonic treatments and experiment is even worse if one looks at the difference between high and low temperature phases.

Since the BKS potential agrees well with experiment when all thermal fluctuation are taken into account via MD simulations and the trend in the harmonic approximation of BKS and *ab initio* are similar, it seems that Demuth et al.'s LDA treatment is quite reasonable and one should acknowledge the fact that these data were made public, albeit the discrepancy between calculations and experiments were pointed out by the authors themselves.

Yet, the message to be learned is that elastic properties and changes in the lattice parameters cannot be obtained from regular *ab initio* calculations in a

Table 1. Lattice constants a and c in α quartz (T = 300 K) and β-quartz (T = 1000 K), along with some representative elastic moduli. Differences between values of a, c, or C_{ij} between the α and the β phase are inserted as well. (a) Carpenter et al. (1998) (experiments), (b) Müser and Binder (2001) (finite T simulations), (c) van Beest, Kramer, and van Santen (1990) (in harmonic approximation, data for T = 1000 K unpublished results from present author), (d) Demuth et al. (1999) (local density approximation, harmonic approximation)

	exper.[a]	MD[b]	BKS (h)[c]	*ab-initio*[d]
a(300 K)	4.914	4.967	4.941	4.899
a(1000 K)	5.000	5.031	5.022	5.026
$100\Delta a/a$	1.72	1.29	1.62	2.59
c(300 K)	5.406	5.470	5.449	5.383
c(1000 K)	5.459	5.528	5.538	5.512
$100\Delta c/c$	0.98	1.06	1.65	2.39
C_{11}(300 K)	086.4	087.1	090.5	
C_{11}(1000 K)	132.2	140.6	266.0	
$\Delta C_{11}/C_{11}$	0.53	0.61	1.96	
C_{33}(300 K)	103.9	101.2	107.0	
C_{33}(1000 K)	121.4	127.5	227.0	
$\Delta C_{33}/C_{33}$	0.17	0.26	1.12	
B_0(300 K)	038.1	036.3	040.5	035.4
B_0(1000 K)	070.2	079.8	164.0	132.6
$\Delta B_0/B_0$	0.84	1.20	3.10	3.75

high-temperature polymorph if the atomic probability distribution functions deviate from Gaussians in a way similar to the deviation seen in β-quartz. Of course Car-Parrinello type simulations allow the determination of all these properties provided that the simulations cells are sufficiently large and the statistics sufficiently good. In the present case, we needed a few 10,000 MD steps and a little less than thousand atoms in order to get elastic constants with an accuracy of about 5 percent accuracy.) While elastic constants can be evaluated much faster in the NVT ensemble than in the NpT ensemble at small temperatures, this statement becomes incorrect in the high-temperature phase, because the thermal fluctuation corrections are in the order of the harmonic constants evaluated in the harmonic approximation. Since time steps are considerably smaller in Car-Parrinello simulations and since each individual time step is much more CPU time expensive, it seems that classical (or path integral) molecular dynamics are the only reliable route to do these calculation for the next few years.

4.3 Nature of the α-β Phase Transition

In the previous discussion, we have only learned that the disorder is non-Gaussian, but nothing has been said to the nature of the disorder. If β-quartz consisted of α_1 and α_2 domains, which spatially and dynamically averaged to the idealized β-quartz structure, then the phase transformation could be expected to be an order-disorder transition. Neutron diffraction [Wright and Lehman 1981], NMR

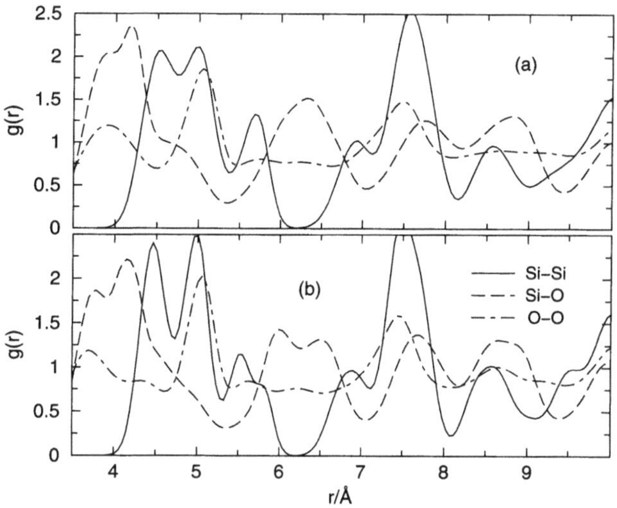

Fig. 9. Radial distribution functions for $3.5\,Å < r < 10.0\,Å$ at temperatures (a) $T =$ 875 K (α-quartz) and (b) $T = 625$ K (β-quartz). From Müser and Binder (2001)

studies [Spearing et al. 1992], and molecular-dynamics simulations [Tsuneyuki et al. 1990] were interpreted as evidence for this scenario. The majority of recent studies, however, favors a displacive type of phase transformation. In this case, the actual structure of the high-temperature phase is interpreted as an ideal β-quartz structure, which is distorted by rigid unit modes of relatively stiff tetrahedral SiO_4 units. This point of view explains the existence of soft modes in the α and β [Axe and Shirane 1970, Tezuka at al. 1991, Carpenter et al. 1998, Dove et al. 1992] and the absence of symmetry forbidden phonons in the β-phase of quartz [Salje et al. 1992]. The non Gaussian behaviour observed by Kihara (1990) mentioned above favored an ordered structure. The non Gaussian behaviour was not interpreted as disorder but as librational motion of the oxygen atoms around the Si-Si lines.

In order to gain insight into the local structure, it is convenient to calculate radial distribution functions $g_{\mathrm{SiSi}}(r)$, $g_{\mathrm{SiO}}(r)$, and $g_{\mathrm{OO}}(r)$. As mentioned above, all these three radial distribution functions can be obtained individually in a straightforward way and with very good accuracy, and thus complement experiments where this information is not easily available. The $g(r)$'s are shown in Fig. 9 a little more than 100 K above and below "our" phase transition temperature of $T_{\mathrm{tr}} = 740\,\mathrm{K} \pm 5\,\mathrm{K}$. (The determination of T_{tr} will be outlined below.) It can be seen that most features are similar in the two phases even for relatively large distances r. This raises the question whether or not one can interpret the $g(r)$'s as measured in β-quartz as a mere temperature broadened version of the α-quartz configuration. If this was the case the picture of the domain disorder in β-quartz would be supported. If it is possible to find qualitative features in

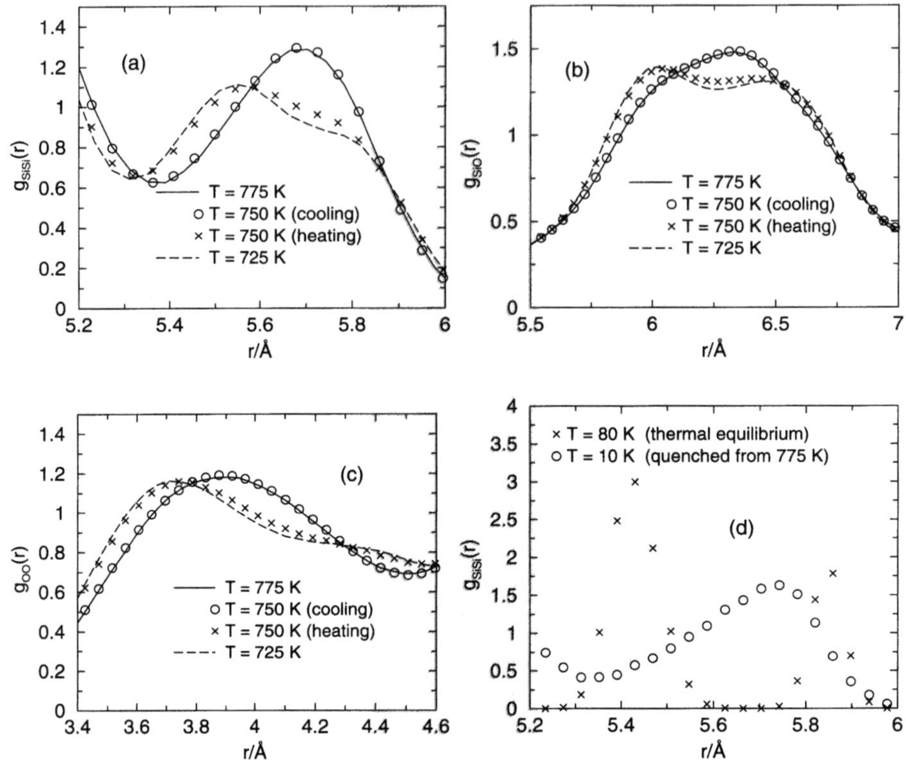

Fig. 10. Details of the radial distribution function Si-Si (a), Si-O (b), and O-O (c) for system size $N = 4320$ at various temperatures. (d) is same as (a), but from configurations that have been quenched from high temperature phase to 10 K.

$g(r)$ in β quartz that are not akin of those in α quartz (in particular for small r), then the domain picture can be ruled out.

In order to investigate this issue in more detail, it is best to calculate the various $g(r)$'s at a temperature where both phases are thermodynamically stable or at least metastable, i.e. in the immediate vicinity of $T_{\rm tr}$. It turns out that all $g(r)$'s evolve rather smoothly in either phase, however, there is a sudden change by going from the α phase to the β phase. This behaviour is illustrated in Fig. 10 for selected areas of various $g(r)$'s. The configurations in Fig. 10 at $T =$ 750 K have been equilibrated for 3000 MD steps before the radial distribution functions were averaged over 2000 MD steps. The initial configurations were equilibrated configurations from 25 K below or above 750 K. Smaller systems, e.g., $N = 1080$ systems relax considerably at $T = 750$ K within the above mentioned equilibration time period. One β quartz configuration was quenched down to $T = 10$ K in order to find the inherent structure of a typical β-quartz configuration. Quenching is done by suddenly dropping the temperature and chosing much larger couplings to the thermostat than usual.

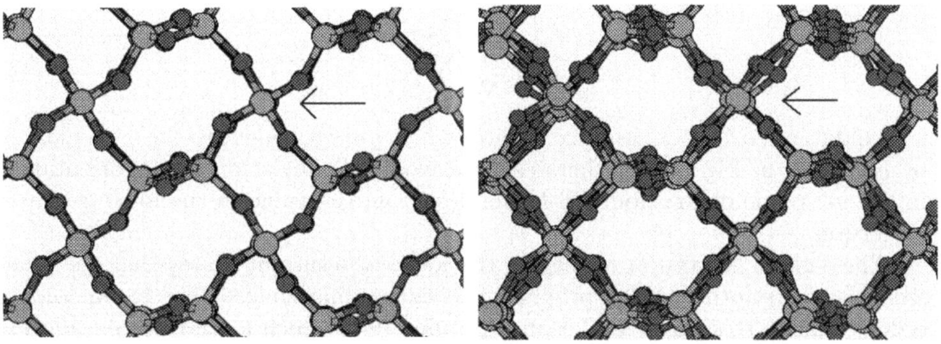

Fig. 11. View along the [100] axis in α-quartz at $T = 80$ K (left) and β-quartz at $T = 1050$ K (right). Dark and light atoms represent oxygen and silicon atoms respectively. Both snapshots belong to identical subvolumes of the simulation cell. The [001] axis goes from the left to the right. The rotation angles about the [100] axis of the units marked by an arrow is used to define the order parameter. From Müser and Binder (2001)

Figure 10 reveals that the local structures in β-quartz can certainly not be interpreted as (temperature) broadened α-quartz domains, e.g., there are clear double peaks in g_{SiSi} at $r \approx 5.6$ Å and g_{SiO} at $r \approx 6.25$ Å in the α phase that are absent in the β phase. The various $g(r)$'s do not change significantly with temperature above T_{tr}, but make sudden changes near and below T_{tr}. The (double) peaks in $g(r)$ become increasingly more pronounced as the temperature is lowered further below T_{tr}. This is furthermore supported by the striking observation that no double peak in the SiSi radial distribution function is observed when $T = 750$ K configurations are quenched down to $T = 10$ K (Fig. 10d). These findings support an experimental study by Tucker et al. (2000), who deduced the nearest-neighbor Si-Si-Si angle distribution from the so-called total pair correlation function $T(r)$. They found that two peaks of the Si-Si-Si angle distribution coalesced upon heating at the $\alpha - \beta$ phase transition.

In order to analyze the α-β transition in more detail, it is instructive to visualize the changes of the structure in quartz. This is done in Fig. 11, where a snapshot along the [100] axis is taken for α-quartz at $T = 80$ K and for β-quartz at $T = 1050$ K. The rotation of tetrahedra about the [100] axis can be seen particularly well for the positions that are equivalent to those sites marked by an arrow. In the β-quartz phase, no α_1 or α_2 domains become apparent. This statement also holds for most configurations obtained near but above T_{tr}. For large system sizes near T_{tr}, it is actually possible to observe jumps of the entire system between configurations that entirely resemble the α-quartz structure and those that resemble β-quartz.

In order to obtain information about the order in the system, a global order parameter ϕ is defined that measures the rotation of (distorted) tetrahedra about the [100] axis, such that

$$\phi = \frac{1}{N^*} \sum_{i^*=1}^{N^*} \varphi_{i^*}, \tag{14}$$

where the sum over i^* is confined to sites which are equivalent to the sites marked by an arrow in Fig. 11. φ_i^* denotes the (averaged) deviation of the orientation in the y-z plane of the (four) Si-O bond(s) from the value in the ideal β-quartz structure.

The sudden structural change in the local structure indicates that the transition is of first order. Yet a proper finite-size scaling analysis has to be done in order to make this sure [Binder and Stauffer 1987]. Such a study could address the speculation whether the first-order nature of the transition merely arises due to the incommensurate instability that occurs at a temperature which is 1.5 K higher than the transition into the α phase. As argued above the incommensurate phase will be strongly suppressed in simulations owing to the large gap between the linear box dimension in our simulation and the length of the wave length of the incommensurate mode.

It is difficult to locate T_{tr} precisely by just calculating the expectation value of the (absolute value) of ϕ, because all thermodynamic properties behave smoothly for finite-size systems near the phase transition temperature. In order to determine the transition temperature T_{tr} nevertheless accurately, use is made of the fourth' order cumulant [Binder 1981], which is defined in the case of a one-component order parameter as

$$g_4(N,T) = \frac{1}{2}\left(3 - \frac{\langle\phi^4\rangle_N}{\langle\phi^2\rangle_N^2}\right), \tag{15}$$

where $\langle\phi^k\rangle_N$ denotes the thermal average of the k'th moment of the order parameter for an N-particle system. It has been shown [Vollmayr et al. 1993] that $g_4(N,T)$, aside from small correction terms, has a size-independent crossing point g_4^* at a first-order phase transition. For the calculation of g_4^*, the geometry of the simulation cell is supposed to be constant. It is difficult to satisfy this requirement without increasing the particle number N considerably for the quartz structure if the cell geometries are approximately cubic. The smallest box length should exceed twice the cutoff radius which limits us to $N \geq 1080$. While it is still possible to equilibrate system sizes of the order $N \approx 2000$ near the phase transition, this becomes extremely difficult for $N \approx 4000$. Note that the equilibration time increases algebraicly with N at a second-order and exponentially with N at a first-order transition point.

The expectation values of the cumulants are shown in Fig. 12. Due to the fact that the cell geometries slightly differ between the $N = 1080$ and the $N = 2160$ system, we cannot expect perfect crossing of the two different systems at T_{tr}. However, comparison to the value where the cumulants cross in a Landau description of this transition (see below for more details) makes it plausible that the crossing of the cumulants shown in Fig. 12 is indeed meaningful. Within the statistical error bars, it is possible to locate the transition at $T_{\text{tr}} \approx 740$ K with an uncertainty of about 5 K for both system sizes.

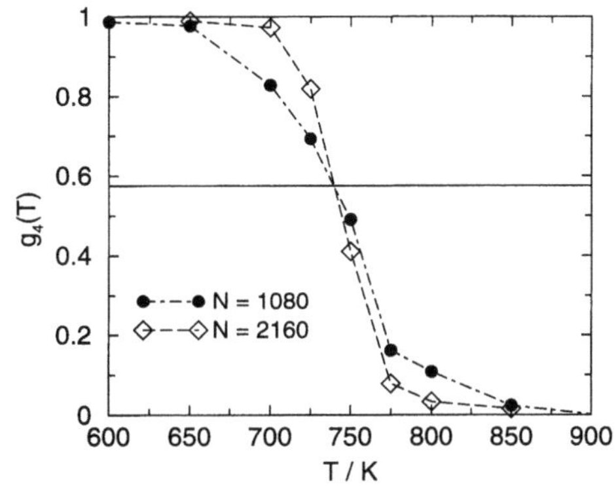

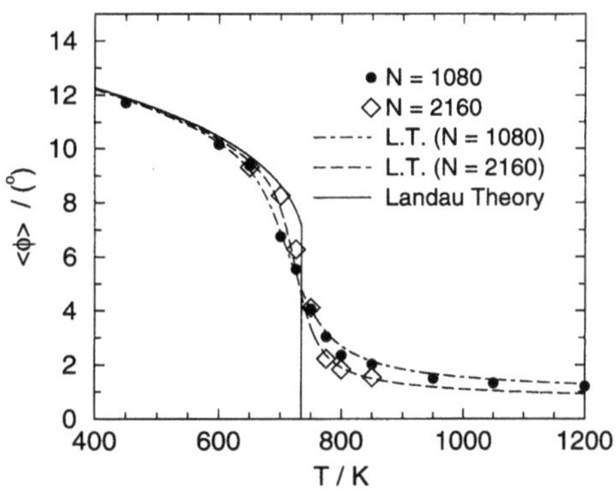

Fig. 12. Left: Fourth order cumulant g_4 as a function of temperature T for two different system sizes. The value g_4^* at which crossing of the cumulants is predicted within Landau theory is indicated by a straight line. Broken lines are drawn to guide the eye. Right: Order parameter $\langle | \phi | \rangle$ as a function of temperature. The lines reflect fits according to Landau theory, whose free parameters were adjusted to the $N = 1080$ curve. The solid line corresponds to the thermodynamic limit in Landau theory. Broken lines represent finite-size Landau theory. From Müser and Binder (2001)

In order to describe the transition within a Landau theory with a single scalar order parameter, we adopt the form [Carpenter et al. 1998]

$$F(\phi, T) = \frac{1}{2}a(T - T_c)\phi^2 + \frac{1}{4}b\phi^4 + \frac{1}{6}c\phi^6, \tag{16}$$

where F is the free-energy per particle as a function of temperature T and order parameter ϕ while a, b, c, and T_c are (free) parameters. In order to find those parameters that are appropriate to describe our simulation results, we need to generalize the approach to *finite* system sizes. This is done by evaluating numerically expectation values of the n'th power of ϕ using ($\beta = 1/k_B T$) in the following way:

$$\langle \phi^n \rangle = \frac{\int_{-\infty}^{\infty} d\phi \, \phi^n \exp\{-\beta N F(\phi, T)\}}{\int_{-\infty}^{\infty} d\phi \exp\{-\beta N F(\phi, T)\}}. \tag{17}$$

The parameters a, b, c, and T_c were determined by fitting the temperature dependence of the order parameter $\langle | \phi | \rangle$ for the $N = 1080$ system. The fit is shown on the right-hand side of Fig. 12 along with similar data for $N = 2160$. We also included data in which the thermodynamic limit was taken. It can be seen that the size effect in $\langle | \phi | \rangle$ is reasonably described by Landau theory.

Overall, similar coefficients as from experimental data are drawn from the simulations. Most importantly, the parameter b is found to be slightly negative, $T_c = 715$ K is obtained about 115 K smaller than in experiment, and $T_t - T_c \approx 20\,K$ is found about twice as large as in real experiment [Carpenter et al. 1998]. Nonetheless, the global picture of experiment is certainly reproduced. Since lattice parameters, elastic constants, and phase transition agree reasonably well with experiment, it is safe to draw qualitative conclusions about properties which are easily accessible in the simulation, but not as easily accessible experimentally, i.e., information on the local structure.

5 Mechanical Stability of Quartz-I and Quartz-II

In this section, we will be concerned with another phase of quartz, namely the high-pressure phase called quartz II. Quartz II is a silica polymorph that can be obtained from α-quartz by slowly increasing the external pressure P to values slightly larger than $P_{\text{I}-\text{II}} = 21$ GPa [Kingma et al. 1993]. The crystal-crystal transition has been observed a little less than a decade ago and the mechanism driving the transition is still subject of a controversial debate [Binggeli and Chelikowsky 1992, Gregoryanz et al. 2000, Müser and Schöffel 2001]. Computer simulations of regular quartz (quartz I) suggested that one of the Born stability criteria (BSC) was expected to be violated in α-quartz at 25 GPa and the transition from quartz I to quartz II was related to this mechanical instability [Binggeli and Chelikowsky 1992]. A reanalysis of the BSC in terms of experimentally measured elastic constants, however, found a large discrepancy to these theoretical predictions [Gregoryanz et al. 2000]: The hypothetical, mechanically (meta)stable pressure regime of quartz I for $P > P_{\text{I}-\text{II}}$ was found to go up to values as large as 40 GPa. Snapshots of the pressure-induced phase transformation are shown in Fig. 13.

The suspicion has been raised that this discrepancy might be due to an inconsistent evaluation of the BSC [Müser and Schöffel 2001]: The definition of second-order elastic constants is not unique at non-zero pressures. It is commonly distinguished between elastic constants C_{ij} and stiffness or Birch coefficients B_{ij} [Wallace 1971]. The latter are obtained from evaluating (7), while the C_{ij} are obtained by using the standard estimators for elastic constants in the NVT ensemble. The proper BSC for non-zero pressures is to require that the matrix B_{ij} be positive definite [Wang et al. 1993]. Thus, when analyzing the mechanical

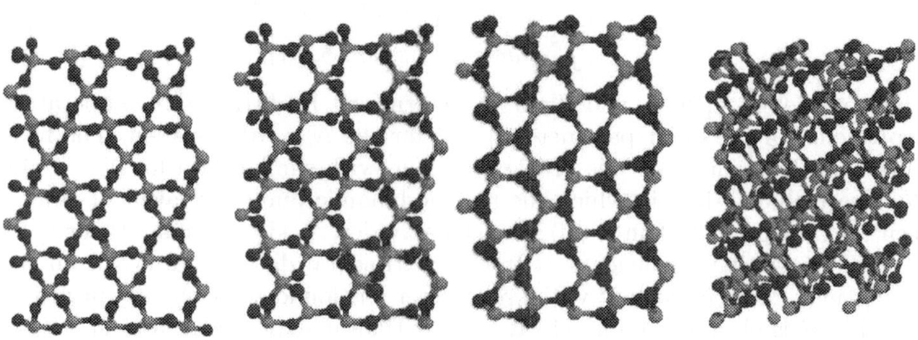

Fig. 13. Snapshots of quartz under pressures at: $p = 0$, 8, 21, and 22 GPa (from the left to the right)

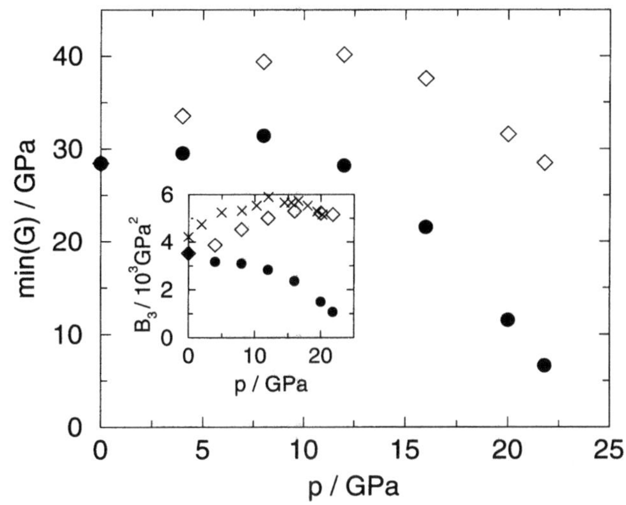

Fig. 14. Minimum eigenvalue of the matrix of the elastic constants (open diamonds) and of the Birch matrix (filled circles) as a function of pressure. Inset: Born stability criterion applied to elastic constants (diamonds) and Birch coefficients (closed circles). Error bars are about symbol size. Crosses represent data from Gregoryanz et al. (2000).

stability of α-quartz under pressure, the BSC

$$B_3 = (C_{11} - C_{12})C_{44} - 2C_{14}^2 > 0 \tag{18}$$

is not meaningful. This condition has often been examined when discussing whether the pressure induced amorphization transition reported in molecular dynamics simulations (MDS) is driven by a mechanical instability [Gregoryanz et al. 2000, Binggeli and Chelikowsky 1992, Tse, Klug and Le Page 1992].

Distinguishing between Birch and elastic constants reveals that the BSC overestimates the stability of α-quartz under pressure considerably. This is shown in Fig. 14. In the inset of that figure it can be seen that the MDS reproduce the experimentally measured pressure dependence of B_3 fairly well. This comparison justifies the use of the employed model potential surface for the stability analysis. If $B_{\alpha\beta}$ replace $C_{\alpha\beta}$ in (18) the behavior is similar to the one reported by Binggeli and Chelikowsky [Binggeli and Chelikowsky 1992], e.g., a zero slope of B_3 at small pressures is found.

The correct stability criterion is examined in the main part of Fig. 14. The smallest eigenvalue of the Birch coefficients tends to zero much faster than the stability criterion (using regular elastic constants) related to (18). From the main part of Fig. 14, mechanical instability would be expected at a pressure $p \approx 25$ GPa, while the former analysis predicts an instability at $p \approx 40$ GPa.

Preliminary analysis shows that the transition between α-quartz and quartz II is kinetically hindered and non-reversible, but these studies will only be published in the near future.

6 Conclusions

These lecture notes gave an overview over selected computer simulation studies of the structure of β-quartz, the temperature driven phase transition between β-quartz and α-quartz and the pressure driven phase transition between α-quartz and quartz II. The relationship between structure and structure fluctuations on one hand and elastic properties and phase transition driving mechanisms on the other hand was explored. Some features turned out to be more complex than those that we know from one-component systems, i.e., the non-Gaussian nature of the local disorder in β-quartz and the consequences for the harmonic properties. Many conclusions drawn here can be believed to be rather generic for solids with similar tetrahedral local order (cristobalite, various phases of GeO_2, etc.).

It was also shown how model potentials for relatively complex systems can be tested in computer simulations. This is done by evaluating lattice parameters, elastic constants, and thermal expansion at low temperature. In the latter case, quantum effects of the ionic motion need to be taken into account. It was also shown that the use of finite-size scaling techniques, which are usually only applied to models (that are computationally much more feasible, Ising model, simple Lennard-Jonesium, etc.) is perfectly adequate to analyze phase transitions in more complex materials.

Acknowledgments

We are grateful for financial support for this research from the Materialwissenschaftliches Forschungszentrum (MWFZ) Mainz, the Bildungsministerium Für Bildung und Forschung (BMBF) in the framework of the Kompetenzzentrum 'Werkstoffsimulation' (grant N° 03N6015) and from Schott Glaswerke. The author thanks K. Binder for useful discussions.

References

[Allen and Tildesley 1987] Allen M. P. and Tildesley D. J. (1987) Computer Simulations of Liquids. (Clarendon Press, Oxford).

[Axe and Shirane 1970] Axe J. C. and Shirane G. (1970) Study of the α-β phase transformation by inelastic bneutron scattering. Phys Rev. B 1: 342-348.

[Barker 1979] Barker J. A. (1979) A quantum-statistical Monte Carlo method; path integrals with boundary conditions, J. Chem. Phys. 70, 2914-2918.

[Binder 1981] Binder K. (1981) Critical properties from Monte Carlo: Coarse graining and renormalization. Phys. Rev. Lett. 47: 693-696.

[Binder and Stauffer 1987] Binder K. and Stauffer D. (1987) in *Applications of the Monte Carlo Method in Statistical Physics*, edited by K. Binder (Springer, Berlin).

[Binggeli and Chelikowsky 1992] Binggeli N. and Chelikowsky J. R. (1992) Elastic instability in α-quartz under pressure. Phys. Rev. Lett. 69, 2220-2223.

[Carpenter et al. 1998] Carpenter MA, Salje EKH, Graeme-Barber A, Wrucki B, Dove MT, and Knight KS (1998) Calibration of excess thermodynamic properties and elastic constant variations associated with the $\alpha \leftrightarrow \beta$ phase transition in quartz. Am. Mineral. 83: 2-22.

[Demuth et al. 1999] Demuth T., Jeanvoine Y., Hafner J., and Angyan J. G. (1999) Polymorphism in silica studied in the local density and generalized-gradient approximation. J. Phys.: Condens. Matter 11: 3833-3874.

[Dolino 1990] Dolino G. (1990) The α-inc-β phase transition of quartz: A century of research on displacive phase transitions. Phase Transitions 21: 59-72.

[Dove et al. 1997] Dove M. T., Kreen D. A., Hannon A. C., and Swainson IP (1997) Direct measurement of the Si-O bond length and orientational disorder in the high-temperature phase of cristobalite. Phys. Chem. Minerals 24: 311-317.

[Dove et al. 1999] Dove M. T., Gambhir M., Heine V. (1999) Anatomy of a structural phase transition: theoretical analysis of the displacive phase transition in quartz and other silicates. Phys. Chem. Minerals 26: 344-353.

[Feynman and Hibbs 1965] Feynman R. P. and Hibbs A. R. (1965) *Quantum Mechanics and Path Integrals.* (Mc.Graw-Hill, New York).

[Frenkel and Smit 1996] Frenkel D. and Smit B. (1996) Understanding Molecular Simulation: From Algorithms to Applications. (Academic Press, San Diego).

[Gregoryanz et al. 2000] Gregoryanz E., Hemley R. J., Mao H. K., Gillet P. (2000) High-pressure elasticity of α-quartz: Instability and ferroelastic transition. Phys. Rev. Lett. 84: 3117-3120.

[Heaney, Prewitt, and Gibbs 1994] Heaney P. J., Prewitt C. T., and Gibbs G. V. (Eds.) (1994) Silica. Physical behavior, geochemistry, and materials applications. (Mineralogical Society of America, Washington, D.C.).

[Kihara 1990] Kihara K. (1990) An X-ray study of the temperature dependence of the quartz structure. Eur. J. Mineral. 2: 63-77.

[Kingma et al. 1993] Kingma K. J., Hemley R. J., Mao H. K., and Veblen D. R. (1993) New high-pressure transformation in α-quartz. Phys. Rev. Lett. 70, 3927-3930.

[Landau and Binder 2000] Landau D.P. and Binder K. (2000) A guide to Monte Carlo simulations in statistical physics. (Cambridge University Press, Cambridge).

[Martonak et al. 1998] Martonak R., Paul W., and Binder K. (1998) Orthorhombic phase of crystalline polyethylene: A constant pressure path-integral Monte Carlo study. Phys. Rev. E 57, 2425-2437 (1998).

[Müser et al. 1995] Müser M. H., P. Nielaba, and K. Binder (1995) Path-integral Monte Carlo of crystalline Lennard-Jones systems, Phys. Rev. B 51, 2723-2731 (1995).

[Müser 2001] Müser M. H. (2001) Simulation of material properties below the Debye temperature: A path-integral molecular dynamics case study of quartz. J. Chem. Phys. 114, 6364-6370.

[Müser and Binder 2001] Müser M. H. and Binder K. (2001) Molecular dynamics study of the α-β transition in quartz: Elastic properties, finite size effects, and hysteresis in the local structure. Phys. Chem. Min. **28**, 746-755.

[Müser and Schöffel 2001] Müser M. H. and Schöffel P. (2001) Comment on: High-pressure elasticity of α-quartz: Instability and ferroelastic transition. cond-mat/0009353.

[Parrinello and Rahman 1980] Parrinello M. and Rahman A. (1980) Crystal structure and pair potentials: A molecular-dynamics study. Phys. Rev. Lett. **45**, 1196-1199.

[Parrinello and Rahman 1981] Parrinello M. and Rahman A. (1981) Polymorphic transitions in single crystals: A new molecular dynamics method. J. Appl. Phys. 52, 7182-7190.

[Parrinello and Rahman 1982] Parrinello M. and Rahman A. (1982) Strain fluctuations and elastic constants. J. Chem. Phys. **76**, 2662-2666.

[Rickwardt et al. 2001] Rickwardt Chr., Nielaba P., Müser, and Binder K. (2001) Path integral Monte Carlo simulation of silicates. Phys. Rev. B 63, 045204 (2001).

[Salje et al. 1992] Salje E. K. H., Ridgwell A., Güttler B., Wruck B., Dove M. T., and Dolino G. (1992). On the displacive character of the phase transition in quartz: A hard mode spectroscopic study. J. Phys.: Condens. Matter 4: 571-577.

[Schneider and Stoll 1978] Schneider T. and Stoll E. (1978) Molecular dynamics study of a three-dimensional one-component model for distortive phase transitions. Phys. Rev. B 17, 1302-1322.

[Schöffel and Müser 2001] Schöffel P. and Müser M. H. (2001), Elastic constants of quantum solids by path integral simulations, Phys. Rev. B 63 (22), 4108-1–4108-9.

[Spearing et al. 1992] Spearing D. R., Farnan I., and Stebbins J. F. (1992) Dynamics of the α-β transition in quartz and cristobalite as observed by in-situ high temperature ^{29}Si and ^{17}O NMR. Phys. Chem. Min. 19: 307-321.

[Striefler and Barsch 1975] Striefler M. E. and G. R. Barsch G. R., Phys. Rev. B **12**, 4553 (1975).

[Tezuka et al. 1991] Tezuka Y., Shin S., and Ishigame M. (1991) Observation of the silent soft phonon in β-quartz by means of hyper-Raman scattering. Phys. Rev. Lett. 66: 2356-2359.

[Tse and Klug 1991] Tse J. S. and Klug D. D. (1991) The structure and dynamics of silica polymorphs using a two-body effective potential. J. Chem. Phys. 95, 9176-9185.

[Tse, Klug and Le Page (1992)] Tse J. S., Klug D. D., and Y. Le Page Y. (1992) High-pressure densification of amorphous silica. Phys. Rev. Lett. **69**, 3647.

[Tsuneyuki et al. 1988] Tsuneyuki S., Tsukada M., Aoki H., and Matsui Y. (1988) First principles interatomic potential of silica applied to molecular dynamics. Phys. Rev. Lett. 61: 869-872.

[Tsuneyuki et al. 1990] Tsuneyuki S., Aoki H., and Tsukada M., (1990) Molecular-dynamics study of the α to β structural phase transition in quartz. Phys. Rev. Lett. 64: 776-779.

[Tucker et al. 2000] Tucker M. G., Dove M. T., Keen, D. A. (2000) Simultaneous analysis of changes in long-range and short-range structural order at the displacive phase transition in quartz. J. Phys.: Condens. Matter 12: L723-L730.

[Tuckerman et al. 1993] Tuckermann M. E., Berne B. J., Martyna G. J., and Klein M. L. (1993) Efficient molecular dynamics and hybrid Monte Carlo algorithms for path integrals. J. Chem. Phys. 99, 2796-2808.

[van Beest, Kramer, and van Santen 1990] van Beest B., Kramer G., and van Santen (1990) Force fields for silicas and aluminophosphates based on ab initio calculations. Phys. Rev. Lett. 64, 1955-1958.

[Vollmayr et al. 1993] Vollmayr K., Reger J. D., Scheucher M., and Binder K. (1993) Finite size effects at thermally-driven first order phase transitions: A phenomenological theory of the order parameter distribution. Z. Phys. B 91: 113.

[Wallace 1971] Wallace D. C., in *Solid State Physics* **25**, ed. H. Ehrenreich, F. Seitz, and D. Turnbull (Academic Press, New York, 1971); *Thermodynamics of Crystals* (Wiley, New York, 1972).

[Wang et al. 1993] Wang J.,Yip S.,Phillpot S. R., and Wolf D. (1993) Crystal instabilities at finite strain. Phys. Rev. Lett. **71**, 4182-4185.

[Wright and Lehman 1981] Wright A. F. and Lehmann M. S. (1981) The structure of quartz at 25 and 590°C determined by neutron diffraction. J. Solid State Chem. 36: 372-380.

Simulation and Theory of Inhomogeneous Liquid Crystals

Mike Allen

Department of Physics and Centre for Scientific Computing,
University of Warwick, Coventry CV4 7AL, United Kingdom

Abstract. This chapter describes Monte Carlo and molecular dynamics simulations, carried out for simple models of liquid crystals. The results are compared with predictions of coarse-grained theories, based on Oseen-Frank orientational elasticity, Landau-de Gennes order tensor theory, and the Onsager density functional. A variety of inhomogeneous systems have been studied, including planar interfaces with containing walls, the equilibrium nematic-isotropic interface, a disclination defect, and a colloidal suspension.

1 Introduction

Liquid crystals may be modelled at various levels of detail. Atomistic simulations, based on empirical potentials or *ab initio* calculations, are intended to calculate the properties of specific molecular systems, but current computer power limits the sample sizes to a few thousand molecules with simulation times up to a few tens of nanoseconds. By using more coarse-grained models, in which the molecule is represented by a single non-spherical unit, these figures may be extended by an order of magnitude or more, at the cost of blurring the connection between molecular structure and liquid crystalline properties. Liquid crystalline properties and phase behaviour are extremely sensitive to details of molecular structure and interactions, so it is a great challenge for either simulation or theory to contribute to the design and synthesis of new materials. Moreover, interest in the liquid crystal community is increasingly focusing on highly inhomogeneous situations (behaviour near interfaces and solid surfaces, domain walls and defects, for example) so some consideration of atomic-scale effects is highly desirable.

In this chapter, we examine the information that can be obtained from simulations which use relatively simple molecular models. In part, the motivation for this comes from the desire to test, and build confidence in, theories of liquid crystals. A theory which is successful for simple models may also apply to more complex models, but greater insight will be obtained by building up the complexity step by step. A second reason to study these simple systems is to develop new ways of measuring, in a simulation, statistical mechanical properties which correspond to experimental measurements. Often there are several different ways of converting the atomic and molecular coordinates into macroscopically relevant results, and one may design different 'computer experiments' to measure the same property. In the search for 'the best' approach (the one subject to the smallest systematic and statistical errors for a given expenditure of computer time) it is most cost-effective to work with the simplest model that contains the essential physics.

The theories that we shall use may all be expressed by writing down a free energy *functional.* The functional may be assumed to depend on simple functions of particle positions and orientations: the local single-particle density, the orientational order tensor, the direction of orientational ordering. All of these descriptions are coarse-grained: specification of the state of the system does not require knowledge of all of the N-particle microscopic coordinates. Instead, we must input the dependence of a handful of functions on the spatial and orientational coordinates, and minimize the free energy with respect to all possible variations of these functions. These very successful continuum models of liquid crystals, based on orientational elastic free energies and hydrodynamics, cover the much larger length- and time-scales of interest to device manufacturers. These theories are (in the main) based on phenomenological coefficients rather than detailed knowledge of molecular structure. Calculating the coefficients may be one aim of simulation; relating them to molecular structure is another.

2 Simulation Models

In our simulation work we use a range of elementary molecular models. Typical examples are *ellipsoids of revolution* of length A, width B; and *spherocylinders* of overall length $L+D$, width D. They are sufficient to describe the basic physics of liquid crystals; they all have a *nematic* phase, which will be the main object of study here. Given these basic shapes, the interaction potentials are divided into 'hard' and 'soft' classes: for *hard* potentials, the energy is infinite when two molecules overlap, and zero otherwise; for *soft* potentials, the energy varies continuously with relative position and orientation. In the work to be described here, we choose a strongly varying, repulsive potential which approximates the hard potential quite well. We may improve these models by including attractive forces [1–5], biaxiality [6,7], nonlinearity [8], flexibility [9,10], dipolar forces [11] and hydrogen bonds [12]. It is also possible to study disk-like shapes, which have their own phase behaviour. However, here we will only be concerned with the simplest rod-like molecular models. The absence of attractions means that the temperature plays a minor role, compared with the density, in determining the thermodynamic behaviour and phase stability.

We use more-or-less standard simulation methods, with some special measures to make the programs run efficiently. Technical details are available elsewhere [13,14], so only a brief summary will be given here. In Metropolis Monte Carlo we use random numbers to translate and rotate the particles randomly, sampling from the canonical, constant-NVT, or isothermal-isobaric, constant-NPT, ensemble. Alternatively, we use constant-NVE and constant-NVT molecular dynamics to advance molecular coordinates according to an equation of motion: the *symplectic* velocity Verlet method is applied to both translation and rotation in our work. Smaller systems (up to tens of thousands of molecules) are typically run on laboratory workstations and PCs, while for the largest systems (up to $N = 10^6$ molecules) a domain decomposition MD program, running on supercomputers such as the Cray T3E at Edinburgh, are used.

3 Theoretical Approaches

Several coarse-grained descriptions of liquid crystals will be of interest. The elastic theory due to Oseen and Frank [15,16] is based on writing part of the free energy $\mathcal{F}$ as a functional of the liquid crystal director $\boldsymbol{n}(\boldsymbol{r})$ which may vary with position $\boldsymbol{r}$:

$$\mathcal{F}[\boldsymbol{n}] = \frac{1}{2}\int_V \mathrm{d}\boldsymbol{r}\; K_{11}\left[\nabla\cdot\boldsymbol{n}\right]^2 + K_{22}\left[\boldsymbol{n}\cdot\nabla\wedge\boldsymbol{n}\right]^2 + K_{33}\left[\boldsymbol{n}\wedge\nabla\wedge\boldsymbol{n}\right]^2 .$$

Here, K_{11}, K_{22} and K_{33} are, respectively, the splay, twist and bend elastic constants. It is possible to add to this an integral over the bounding surface, representing the coupling between the liquid crystal director and the container walls: this so-called *anchoring* free energy will be of interest later.

The Frank free energy describes a nematic liquid crystal in regions where the *direction* of ordering is the only important variable, and where, moreover, it varies smoothly with position (hence it takes the form of a gradient expansion). A theory which allows the *degree* as well as the direction of ordering to vary is due to Landau and de Gennes [17], and is based on the 2nd-rank order tensor $\mathsf{Q}(\boldsymbol{r})$:

$$\mathcal{F}[\mathsf{Q}] = \int_V \mathrm{d}\boldsymbol{r}\; \kappa\left|\nabla\mathsf{Q}\right|^2 + a\mathrm{Tr}\left[\mathsf{Q}^2\right] - b\mathrm{Tr}\left[\mathsf{Q}^3\right] + c\left\{\mathrm{Tr}\left[\mathsf{Q}^2\right]\right\}^2 .$$

On the right, we see a one-constant representation of elastic behaviour (the squared gradient term), plus the leading terms in a free energy expansion, which would give a phenomenological description of the bulk I-N transition in the absence of gradients.

The final theory to be considered in this chapter is that of Onsager [18] in which the free energy, is written as a functional of the single-particle density $\varrho^{(1)}(\boldsymbol{r},\boldsymbol{u})$:

$$\begin{aligned}\mathcal{F}[\varrho^{(1)}] = k_{\mathrm{B}}T\int \mathrm{d}\mathbf{1}\; \varrho^{(1)}(\mathbf{1})\left[\ln\varrho^{(1)}(\mathbf{1})\Lambda^3 - 1\right] \\ -\frac{1}{2}k_{\mathrm{B}}T\int\!\!\int \mathrm{d}\mathbf{1}\mathrm{d}\mathbf{2}\; \varrho^{(1)}(\mathbf{1})\varrho^{(1)}(\mathbf{2})\; f(\mathbf{1},\mathbf{2}) .\end{aligned} \tag{1}$$

Here, $\boldsymbol{r}$ is the position and $\boldsymbol{u}$ the orientation vector; we abbreviate $(\boldsymbol{r}_1,\boldsymbol{u}_1)\equiv\mathbf{1}$, $(\boldsymbol{r}_2,\boldsymbol{u}_2)\equiv\mathbf{2}$. Equation (1) takes the form of the first two terms in a virial expansion. The first term represents the free energy of an ideal gas mixture, where Λ is the thermal de Broglie wavelength. In the second virial term,

$$f(\mathbf{1},\mathbf{2}) = f(\boldsymbol{r}_{21},\boldsymbol{u}_1,\boldsymbol{u}_2) = \mathrm{e}^{-v(\boldsymbol{r}_{21},\boldsymbol{u}_1,\boldsymbol{u}_2)/k_{\mathrm{B}}T} - 1 ,$$

is the Mayer function; $v(\boldsymbol{r}_{21}, \boldsymbol{u}_1, \boldsymbol{u}_2)$ is the pair potential, which is taken to depend on the separation vector $\boldsymbol{r}_{21} = \boldsymbol{r}_2 - \boldsymbol{r}_1$ and the orientations. It is in this term that the details of molecular shape and structure enter: indeed, one can apply the theory to exactly the molecular models that are used in a simulation, without introducing any adjustable parameters, and this is the approach that we shall use. The isotropic-nematic transition occurs at low density for highly elongated molecules, so in this limit Onsager theory is quite accurate, at least for bulk properties; part of our interest will be in the possibilities of extending this success to inhomogeneous situations.

Onsager's theory is a fore-runner of what we now call density functional theories. These have proved extremely successful in recent years when applied to simple atomic fluids [19–21], and have been used to explore liquid crystal phase diagrams [22]. In general, the functional $\mathcal{F}[\varrho^{(1)}]$ is unknown and nonlocal; the above, explicit, expression is valid for low densities, but note that it is not restricted to smooth variations of $\varrho^{(1)}(\boldsymbol{r}, \boldsymbol{u})$. The aim in Onsager's theory, as in most density functional theories, is to minimize the grand potential

$$\Omega = \mathcal{F} - \mu N = \mathcal{F} - \mu \int \mathrm{d}\mathbf{1}\varrho^{(1)}(\mathbf{1})$$

by varying the single particle density $\varrho^{(1)}(\boldsymbol{r}, \boldsymbol{u})$ at fixed chemical potential μ. In practice, this is done as follows. Suppose that we only consider situations in which the *spatial* variation of the density occurs in the z direction: the system is assumed to be translationally invariant in the x and y directions. We tabulate $\varrho^{(1)}$ on a grid:

$$\varrho^{(1)}(z_i, \boldsymbol{u}) = \exp\left\{\sum_{\ell m} C_{i;\ell m} Y_{\ell m}(\boldsymbol{u})\right\} .$$

The $Y_{\ell m}(\boldsymbol{u})$ are spherical harmonic functions; typically we expand the density in coefficients $C_{i;\ell m}$ up to order $\ell_{\max} = 4$. Putting the expansion in an exponent (i.e. expanding the logarithm of $\varrho^{(1)}$ rather than $\varrho^{(1)}$ itself) has some technical advantages: for instance, $\varrho^{(1)}$ can never have negative values. Next, we pre-calculate the x- and y-integrations involving the Mayer function, defining an 'excluded area' at prescribed orientation and z-separation

$$\mathcal{A}(z_{21}, \boldsymbol{u}_1, \boldsymbol{u}_2) = -\int \mathrm{d}x_{21}\mathrm{d}y_{21} f(\boldsymbol{r}_{21}, \boldsymbol{u}_1, \boldsymbol{u}_2) .$$

This function may vary rapidly with $\boldsymbol{u}_1$ and $\boldsymbol{u}_2$: typically it is expanded in spherical harmonics up to $\ell_{\max} = 10$. All of the expansion coefficients are tabulated on a grid in z, with a spacing $\delta z \approx 0.01A$ where A is the molecule length. Then the grand potential is calculated by numerical quadrature and minimized by the conjugate gradient method [23].

4 Surface Anchoring Coefficient

Our first example [24,25] is the use of simulation and theory to explore the anchoring of a liquid crystal at a solid surface. To begin with, we consider a sim-

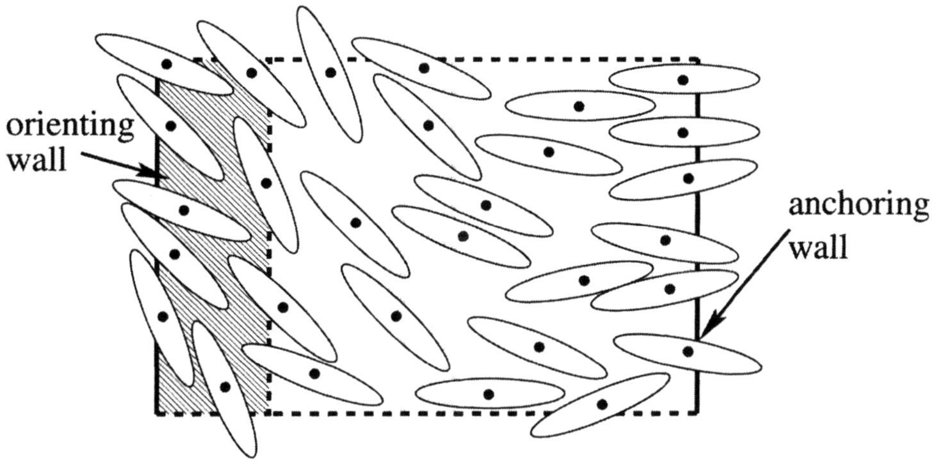

Fig. 1. Schematic picture of simulation to measure wall anchoring coefficient W. At the left wall, a strong, but short-ranged, orienting field is applied in the shaded region to fix the nematic director in a chosen direction. The director profile is measured across the cell: in the elastic theory, it is determined by the surface anchoring coefficient at the right wall, W, and the bulk splay and bend elastic constants K_{11}, K_{33}.

ple 'experimental' geometry, in which the liquid crystal is confined between two walls: at one wall an orienting field is applied, and the response of the director alignment across the cell is measured, with the aim of calculating the strength of anchoring at the opposite wall. In keeping with the simple molecular model, namely hard ellipsoids of elongation $A/B = 15$, the wall-molecule interaction is a very idealized hard potential, defined simply by the condition that the centres of the molecules may not pass through it. Purely entropic considerations, having their origin in the way molecules pack together, generate homeotropic (normal) anchoring. Experimentally, it is quite common to anchor liquid crystals at a treated interface, in which a layer of organic molecules is adsorbed on the solid surface. The liquid crystal molecules partially penetrate the organic layer: the effect of this is modelled (albeit in a simple way) by the chosen boundary condition.

The wall anchoring coefficient W measures the strength with which the surface resists director deformation away from the preferred direction. The approach used in the simulation to measure this is illustrated in Fig. 1. The key quantity turns out to be the *extrapolation length* $\lambda = K_{33}/W$, measurable by fitting director profiles to the elastic theory. For $A/B = 15$ ellipsoids, λ is of the order of one molecular length, as shown in Fig. 2. Direct minimization of the Onsager free energy functional, with essentially no adjustable parameters, reproduces these director profiles very well, *even in the wall regions where the elastic theory is inaccurate.*

In the above treatment, it was assumed that the positions of the elastic-theory boundary conditions coincide with the hard potential walls. There is no

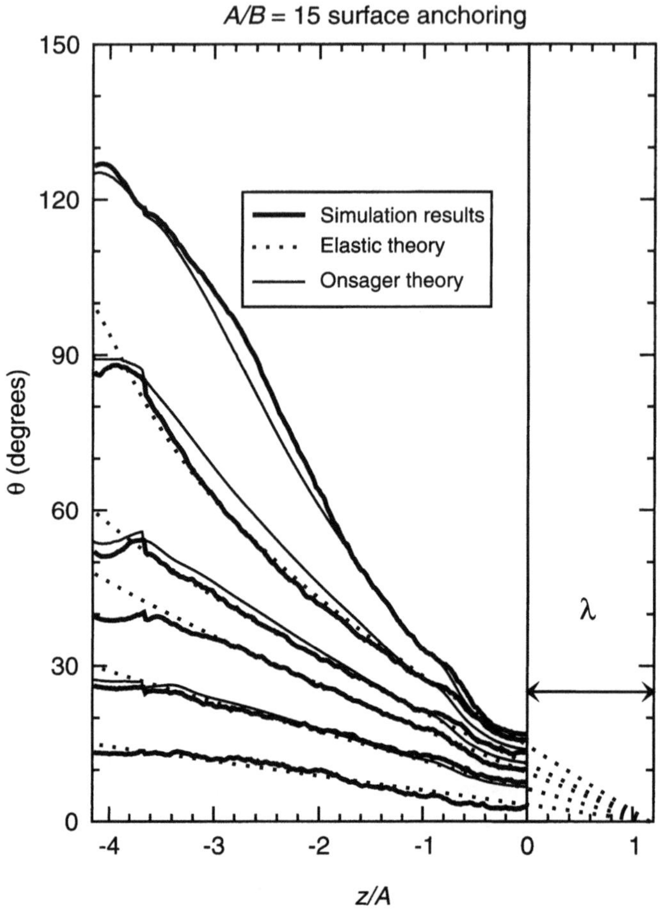

Fig. 2. Simulation results, elastic theory fits, and predictions of Onsager theory to determine the extrapolation length λ, and hence the surface anchoring coefficient W. Several different anchoring angles are imposed at the left wall. The curvature in the profiles is because the deformation is predominantly splay as $\theta \to 90°$ and predominantly bend as $\theta \to 0°$. After [24]

reason, in principle, why this should be so, although we would expect the difference in positions to be a molecular length rather than a mesoscopic length. This difference turns out to be detectable when one studies fluctuations of the director $\boldsymbol{n}(\boldsymbol{r})$ in slabs of varying thickness between identical walls. The fluctuation amplitudes may be expanded:

$$\delta\boldsymbol{n}(\boldsymbol{r}) = \frac{1}{V}\sum_{\boldsymbol{q}_\perp, q_z} \mathrm{e}^{\mathrm{i}\boldsymbol{q}_\perp\cdot\boldsymbol{r}_\perp}\left[\delta\boldsymbol{n}^{(+)}(\boldsymbol{q}_\perp, q_z)\mathrm{e}^{\mathrm{i}q_z r_z} + \delta\boldsymbol{n}^{(-)}(\boldsymbol{q}_\perp, q_z)\mathrm{e}^{-\mathrm{i}q_z r_z}\right] .$$

where $\boldsymbol{q}_\perp = (q_x, q_y)$ is a transverse wave vector, and q_z a wavenumber in the direction normal to the walls. The elastic theory gives a prediction for this ex-

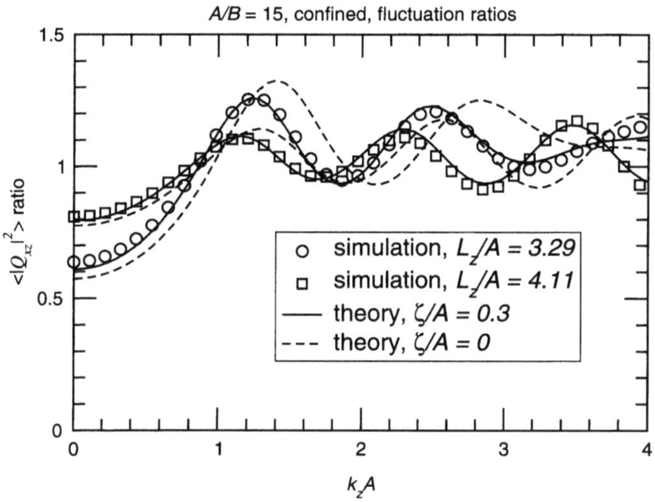

Fig. 3. Squared director fluctuation amplitudes, as functions of wavenumber, for wall separations $L_z/A = 3.29, 4.11$, both normalized by the corresponding results for $L_z/A = 4.93$, to accentuate the characteristic resonances. The predictions of elastic theory, on the assumption $L = L_z + 2\zeta$ with $\zeta = 0$, $0.3A$ are shown. After [26]

pression which is parametrized by the dimensionless quantities $\chi = q_z L$, and $\xi = WL/K = L/\lambda$, where λ is the extrapolation length and L the separation of the boundaries at which the anchoring condition is assumed to apply. This confinement in the z direction, with finite anchoring energy W, gives an *uneven* discrete spectrum q_z, which in turn provides a route to L, independent of any knowledge of the physical wall positions. In simulations, fluctuations of $\boldsymbol{n}(\boldsymbol{r})$ are measured through long-wavelength components of the order tensor $\mathrm{Q}_{\alpha z}(\boldsymbol{r}) = (3/2)Q\delta n_\alpha(\boldsymbol{r})$ where Q is the order parameter. Using the same molecular model as before, director fluctuations between parallel hard walls, with homeotropic anchoring, separated by $L_z/A = 3.29, 4.11, 4.93$, have been investigated [26]. Setting $L = L_z + 2\zeta$, it may be assumed that ζ is characteristic of the wall, independent of L_z. The ratios

$$\frac{\left\langle |Q_{\alpha z}(k_z, L_z)|^2 \right\rangle}{\left\langle |Q_{\alpha z}(k_z, L_z')|^2 \right\rangle}$$

turn out to be very sensitive to ζ, as indicated in Fig. 3. This allows us to separate, in practice, the dependence of the fluctuation amplitudes on W and ζ. The data is clearly incompatible with $\zeta = 0$, and is best fitted with $\zeta/A \approx 0.3$. For this model, the anchoring coefficient varies in an interesting way with density or, equivalently, nematic order parameter. From the available experimental data on thermotropic liquid crystals, one can say that the extrapolation length $\lambda = K_{33}/W$ is inversely proportional to the squared value of the bulk order parameter $\lambda \propto Q^{-2}$. Taking into account that the elastic constant K_{33} is typi-

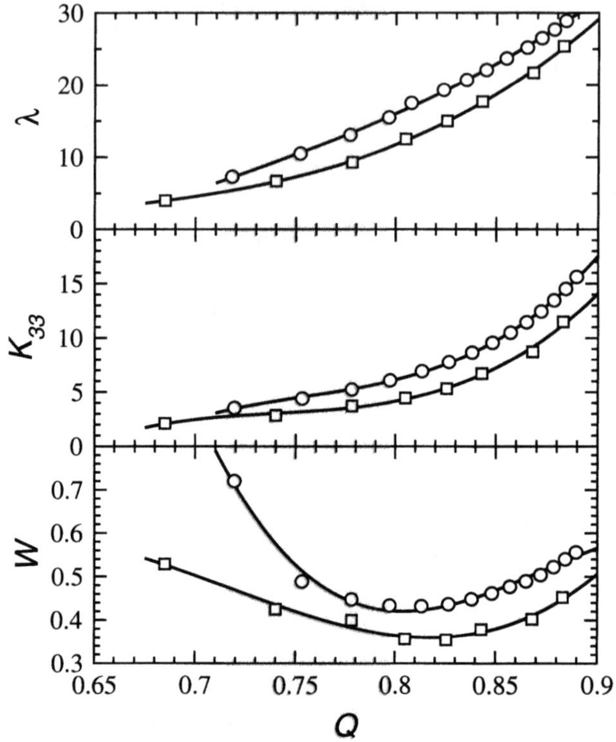

Fig. 4. Extrapolation length λ, bend elastic constant K_{33} and anchoring strength W, as functions of order parameter Q. We show the results of Onsager theory (circles) and Monte Carlo simulations (squares). After [27]

cally proportional to Q^2, this gives for the anchoring parameter $W \propto Q^4$ [28]. However, our simulations and Onsager theory do not conform to this model. The dependences of λ, K_{33}, and W on the order parameter Q are shown in Fig 4. Rather than decreasing, the extrapolation length λ *increases* with Q, albeit not as strongly as K_{33}. The net result is that $W = K_{33}/\lambda$ has a non-monotonic dependence on Q, first decreasing, then increasing, although the variation is not dramatic. It turns out that this behaviour is connected with the variation of the density profile in the immediate vicinity of the surface, which is often neglected in theoretical treatments. Further details may be found in [27].

5 Nematic-Isotropic Interface

Now we consider what happens when the nematic and isotropic phases are at coexistence, either with or without the confining effect of walls. The N-I interface is one of the simplest fluid-fluid interfaces and poses some interesting basic questions. The simulation of this interface requires great care: the coexisting densities are typically very similar, and the surface tension will be relatively small,

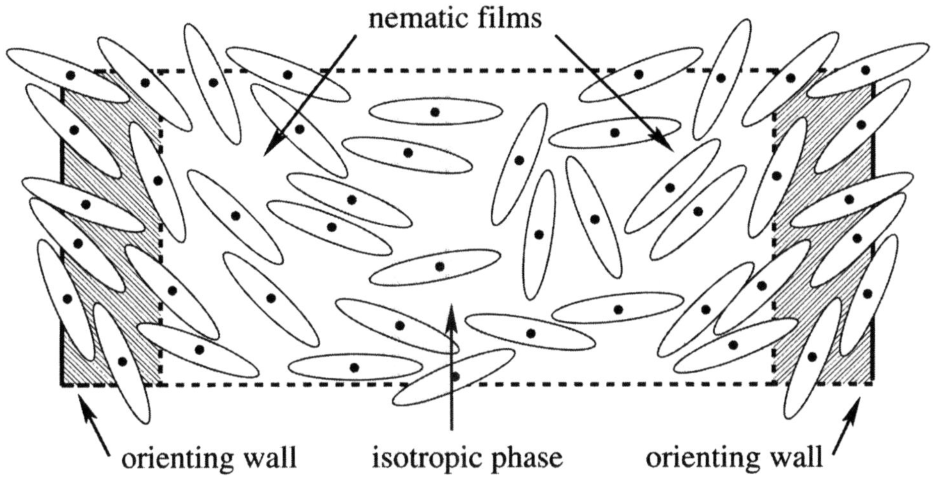

Fig. 5. Schematic picture of simulation to investigate the isotropic-nematic interface, with director orientation controlled by anchoring at the confining walls. At the hard walls, strong, but short-ranged, orienting fields are applied to fix the nematic director in a chosen direction. The two wall anchoring conditions may be chosen independently, provided the isotropic region in the middle is sufficiently wide.

so fluctuations may be important. Our interest here lies in the determination of the properties of the nematic-isotropic interface, including the surface tension, while taking care to allow all the fluctuations associated with thermal equilibrium. We have studied [29] a thick film of nematic phase adsorbed on a solid wall with the same kind of boundary conditions discussed earlier; the system is illustrated schematically in Fig. 5.

5.1 Orientational Order and Density Profiles

For an isolated surface, the film thickness is determined by the deviation of the chemical potential from the bulk coexistence value $\mu - \mu_{\mathrm{IN}}$. In a closed, confined system with adsorbed films on both walls, assuming the wall separation is large enough that the two films are essentially independent, the film thickness is determined by the overall density. The phase transitions in this type of system, such as the wetting behaviour, and the nematic-isotropic equivalent of 'capillary condensation', are of great interest [30,31]; but here the focus of attention will be the structure of the nematic film and the interface between the two fluid phases.

We have, once more, studied ellipsoids of molecular elongation $A/B = 15$. Figure 6 gives an example of the order parameter and density profiles obtained by simulation, and compared with Onsager theory predictions, for the case when the orienting field at the hard wall aligns the director in the plane of the wall, i.e. at 90° to the interface normal. The agreement between theory and simulation is good, bearing in mind that Onsager theory is known to overestimate the bulk coexistence densities; indeed, very detailed comparisons of the profile shapes may be made [29].

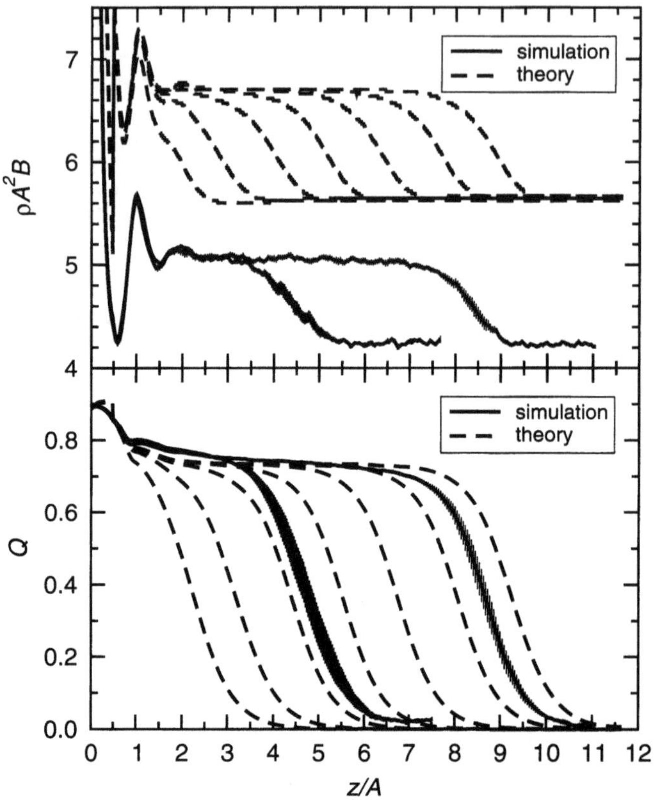

Fig. 6. Profiles of number density in reduced units $\rho A^2 B = NA^2 B/V$, and order parameter Q (largest eigenvalue of the order tensor Q), for a nematic film anchored at 90° to the normal at the left wall (in-plane alignment), in equilibrium with the isotropic phase. Simulation results are given for two different film thicknesses (solid line with error bars) and Onsager theory predictions for seven different film thicknesses (dashed lines). After [29]

There is some interest in studying even longer molecules [32]. Firstly, it has been predicted using Onsager theory for infinitely long rods [33] that the number density $\rho(z)$ varies in a significantly non-monotonic way near the I-N interface, for the lowest free energy case of in-plane alignment. There is no evidence of this for $A/B = 15$, in either simulation or theory (see Fig. 6), which raises the possibility that the effect is only seen for higher elongations. Secondly, again for this alignment, the orientational ordering near the interface should be *biaxial*: symmetry is broken in two directions, namely normal to the interface, and in the plane of the interface along the director. However, this biaxiality is extremely small for the $A/B = 15$ elongation described earlier. In Fig. 7 are the relevant profiles for a soft spherocylinder model with length/width = $L/D = 50$, studied by molecular dynamics, using a free-standing nematic film surrounded by isotropic phase in fully periodic boundaries. It can be seen that the density

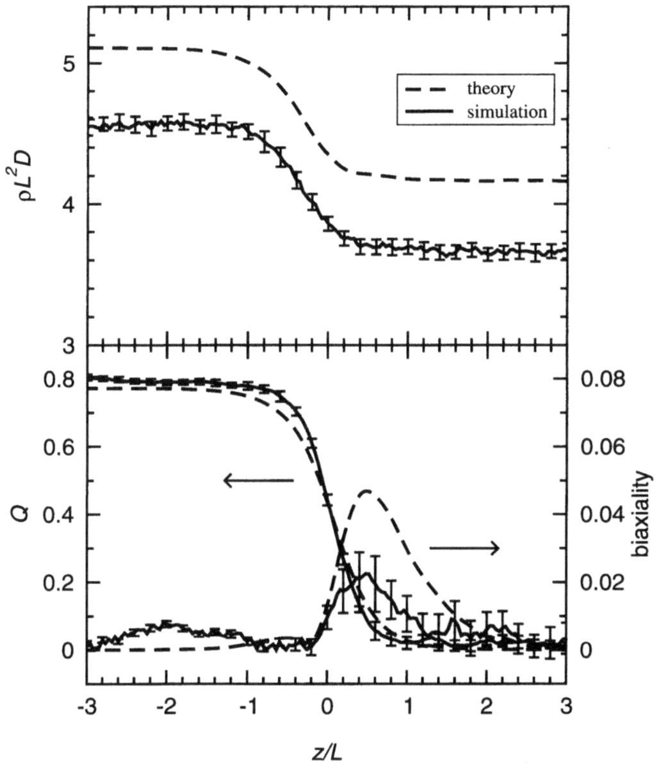

Fig. 7. Profiles of reduced number density $\rho L^2 D$, orientational order parameter Q, and orientational biaxiality (on a $\times 10$ expanded scale), in the vicinity of the I-N interface, for $L/D = 50$ soft spherocylinders, determined by simulation (solid lines with error bars) and from Onsager theory (dashed lines). After [32]

profile changes monotonically, both in simulation and theory. However, the biaxiality (defined as the difference between the middle and lowest eigenvalues of the order tensor Q) is detectable in the simulation, and is overestimated by the Onsager theory by about a factor of 2.

5.2 Surface Tension

One of the most important characteristics of an interface is the surface tension or surface free energy γ. For the I-N interface, the variation of this quantity with director angle gives, essentially, the anchoring potential at the interface. It will dictate the shape of droplets of one phase suspended in the other. Determining this quantity for an equilibrium I-N interface is a significant challenge for computer simulation. For a planar interface normal to the z-direction, a microscopic expression for surface tension γ is:

$$\gamma = \int_{-\infty}^{\infty} \mathrm{d}z \; \left[P_{\|} - P_{\perp}(z)\right] \equiv \int_{-\infty}^{\infty} \mathrm{d}z \; \pi(z) \, .$$

Here, $P_{\parallel} = P_{zz} = P$ is the normal component of the pressure tensor, which is independent of z throughout the system; $P_{\perp}(z) = P_{xx}(z) = P_{yy}(z)$ is the transverse component of the pressure tensor, and this varies with z near interfaces. Far from the interface, $P_{\perp}(z) = P_{\parallel} = P$. In the above formulae, the microscopic definition of $P_{\perp}(z)$ is not unique, although the value of the surface tension does not depend on the choice. In planar geometry, the various valid definitions of transverse pressure correspond to different choices of a contour between two points in the fluid [34]. The two most common conventions, used in simulations, are those of Harasima [35] and of Irving and Kirkwood [36]:

$$\begin{aligned} P_{xx}^{\mathrm{H}}(z_1) &= k_{\mathrm{B}}T\rho(z_1) - \frac{1}{2}\int \mathrm{d}\boldsymbol{r}_{21}\int \mathrm{d}\boldsymbol{u}_1\int \mathrm{d}\boldsymbol{u}_2 \left(x_{21}\frac{\partial v_{21}}{\partial x_{21}}\right) \\ &\quad \times \varrho^{(2)}(\boldsymbol{r}_{21}, z_1, z_1 + z_{21}; \boldsymbol{u}_1, \boldsymbol{u}_2) \\ P_{xx}^{\mathrm{IK}}(z_1) &= k_{\mathrm{B}}T\rho(z_1) - \frac{1}{2}\int \mathrm{d}\boldsymbol{r}_{21}\int \mathrm{d}\boldsymbol{u}_1\int \mathrm{d}\boldsymbol{u}_2 \left(x_{21}\frac{\partial v_{21}}{\partial x_{21}}\right) \\ &\quad \times \int_0^1 \mathrm{d}\zeta\, \varrho^{(2)}(\boldsymbol{r}_{21}, z_1 - \zeta z_{21}, z_1 + (1-\zeta)z_{21}; \boldsymbol{u}_1, \boldsymbol{u}_2)\,. \end{aligned}$$

Here $\varrho^{(2)}(\boldsymbol{r}_{21}, z_1, z_2; \boldsymbol{u}_1, \boldsymbol{u}_2)$ is the pair density and v_{21} the pair potential. These expressions are suitable for evaluation in a molecular dynamics or Monte Carlo simulation. For soft potentials, the derivative $\partial v_{21}/\partial x_{21}$ is just the negative of a component of the intermolecular force. For hard particles the forces are impulsive in nature, and cannot be straightforwardly evaluated in a Monte Carlo simulation; nonetheless, the appropriate contribution to the average pressure tensor may be evaluated by a scaling trick [37,38]. The above formulae are not ideal for use in Onsager's theory, however; in fact, in most density functional theories, the surface tension is calculated directly from the grand potential density, not from the pressure tensor. However, it turns out to be possible to cast the above equations into a suitable form by inserting the low-density form for the pair density

$$\varrho^{(2)}(\boldsymbol{r}_{21}, z_1, z_2; \boldsymbol{u}_1, \boldsymbol{u}_2) \approx \varrho^{(1)}(z_1, \boldsymbol{u}_1)\varrho^{(1)}(z_2, \boldsymbol{u}_2)\exp(-v_{21}/k_{\mathrm{B}}T)\,,$$

and performing a partial integration with respect to x_{21}:

$$\begin{aligned} P_{\perp}^{\mathrm{H}}(z_1) &= k_{\mathrm{B}}T\rho(z_1) - \frac{1}{2}k_{\mathrm{B}}T\int \mathrm{d}\boldsymbol{r}_{21}\int \mathrm{d}\boldsymbol{u}_1\int \mathrm{d}\boldsymbol{u}_2\, f(\boldsymbol{r}_{21}, \boldsymbol{u}_1, \boldsymbol{u}_2) \\ &\quad \times \varrho^{(1)}(z_1, \boldsymbol{u}_1)\varrho^{(1)}(z_1 + z_{21}, \boldsymbol{u}_2) \\ P_{\perp}^{\mathrm{IK}}(z_1) &= k_{\mathrm{B}}T\rho(z_1) - \frac{1}{2}k_{\mathrm{B}}T\int \mathrm{d}\boldsymbol{r}_{21}\int \mathrm{d}\boldsymbol{u}_1\int \mathrm{d}\boldsymbol{u}_2\, f(\boldsymbol{r}_{21}, \boldsymbol{u}_1, \boldsymbol{u}_2) \\ &\quad \times \int_0^1 \mathrm{d}\zeta\, \varrho^{(1)}(z_1 - \zeta z_{21}, \boldsymbol{u}_1)\varrho^{(1)}(z_1 + (1-\zeta)z_{21}, \boldsymbol{u}_2)\,. \end{aligned}$$

The Mayer function f is part of the input to the Onsager minimization, while the equilibrium densities $\varrho^{(1)}(z, \boldsymbol{u})$ are available once the grand potential has been minimized.

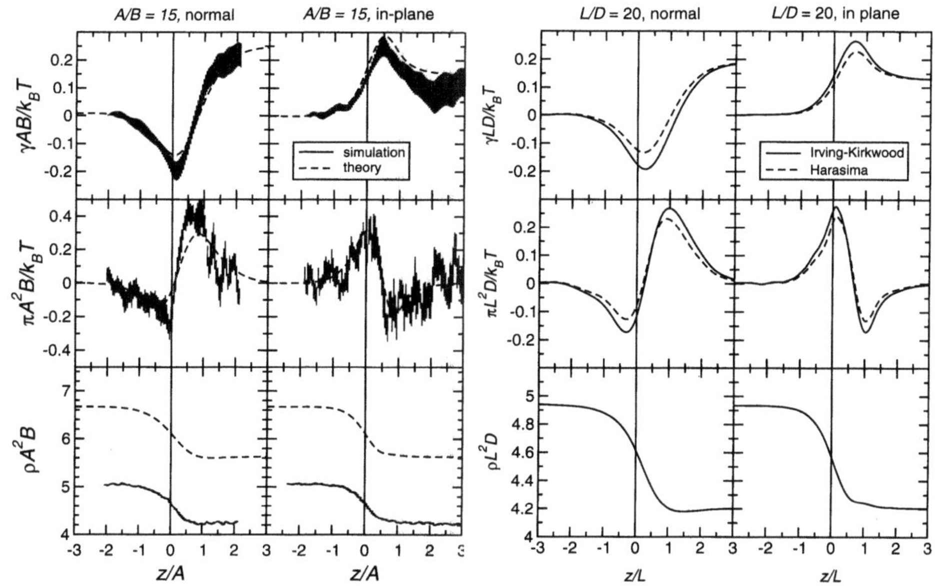

Fig. 8. Profiles of number density ρ, pressure tensor difference $\pi(z)$, and running integral $\int \pi(z)\,dz$ (which gives the surface tension γ), all in appropriate reduced units. Director alignment is normal to the interface or in the plane of the interface, as indicated. **Left:** $A/B = 15$ hard ellipsoids. Simulation results are represented by error bars, calculated from 8 independent simulations, and the predictions of Onsager theory are represented by dashed lines. **Right:** $L/D = 20$ soft spherocylinders, calculated using Onsager theory. The Irving-Kirkwood convention is represented by full lines, the Harasima convention by dashed lines. After Refs [39,40]

The pressure tensor profiles for the I-N interface of hard and soft ellipsoids with $A/B = 15$ are measurable in simulation, taking care to eliminate systematic errors, finite size effects, and statistical fluctuations [39]. The results are shown in Fig. 8. Several interesting things emerge. The simulation results are quite noisy, but they are confirmed in detail by the Onsager theory; the surface tension of the in-plane orientation is indeed found to be lower than for normal alignment; and most dramatically, the curves for the two director orientations are qualitatively different in character. For in-plane alignment, the tension appears on the nematic side of the interface, and there is a small compressive region on the isotropic side, reducing the eventual value of γ somewhat. The net effect is that the surface of tension, defined by the first moment of the pressure tensor profile, is about one molecule length away from the interface on the nematic side. For normal alignment, the tension appears on the *isotropic* side, and there is some compression on the nematic side; the surface of tension is a significant distance, about two molecule lengths, from the interface on the isotropic side.

As mentioned above, one should not read too much into conclusions regarding the pressure tensor profiles themselves, because they depend on the microscopic convention adopted for $P_\perp$, even though the physically measurable surface tension γ does not. Onsager theory can shed some light on the extent of this dependence [40]. For $L/D = 20$ spherocylinders, typical profiles are shown in Fig. 8. Although there is a clear dependence on the choice of Irving-Kirkwood or Harasima convention, the effect is quite small compared with the dramatic consequences of varying the director orientation. Consequently, in this case, we may conclude that this is a genuine physical effect.

In the Onsager approximation, the link between the pressure tensor route to γ and the route based on the grand potential density is clarified. This is at the heart of the thermodynamic definition of the surface tension:

$$\gamma = \int \mathrm{d}z \; [P_{\parallel} - P_{\perp}(z)] = \int \mathrm{d}z \; [P + \omega(z)] = (\Omega - \Omega_{\mathrm{bulk}})/\mathcal{A}\,;$$

$\omega(z)$ is a density which satisfies $\Omega = \int \mathrm{d}\boldsymbol{r}\; \omega(z) = \mathcal{A}\int \mathrm{d}z\; \omega(z)$ where $\mathcal{A}$ is the cross-sectional area. The key to the connection is provided by the condition that the functional derivative of Ω vanishes at equilibrium:

$$\begin{aligned}\frac{\delta\Omega\left[\varrho^{(1)}\right]}{\delta\varrho^{(1)}(\boldsymbol{r}_1,\boldsymbol{u}_1)} &= k_{\mathrm{B}}T\ln\varrho^{(1)}(\boldsymbol{r}_1,\boldsymbol{u}_1)\Lambda^3 - \mu \\ &\quad - k_{\mathrm{B}}T\int \mathrm{d}\boldsymbol{r}_{12}\int \mathrm{d}\boldsymbol{u}_2\; \varrho^{(1)}(\boldsymbol{r}_1+\boldsymbol{r}_{12},\boldsymbol{u}_2)f(\boldsymbol{r}_{12},\boldsymbol{u}_1,\boldsymbol{u}_2) \\ &= 0\end{aligned}$$

This relation may be used to cast the Harasima pressure tensor into a form of $\omega(z)$ which is immediately recognizable in (1).

$$\begin{aligned}&P_{\perp}^{\mathrm{H}}(z_1) = \\ &-k_{\mathrm{B}}T\int \mathrm{d}\boldsymbol{u}_1\; \varrho^{(1)}(z_1,\boldsymbol{u}_1)\left[\ln\varrho^{(1)}(z_1,\boldsymbol{u}_1)\Lambda^3 - 1 - \mu/k_{\mathrm{B}}T\right] \\ &+\frac{1}{2}k_{\mathrm{B}}T\int \mathrm{d}\boldsymbol{r}_{21}\int \mathrm{d}\boldsymbol{u}_1\int \mathrm{d}\boldsymbol{u}_2\; \varrho^{(1)}(z_1,\boldsymbol{u}_1)\varrho^{(1)}(z_1+z_{21},\boldsymbol{u}_2)\; f(\boldsymbol{r}_{21},\boldsymbol{u}_1,\boldsymbol{u}_2) \\ &\equiv -\omega^{\mathrm{H}}(z_1)\,.\end{aligned}$$

Like $P_{\perp}(z)$, the definition of $\omega(z)$ is non-unique; a similar manipulation of the expression for $P_{\perp}^{\mathrm{IK}}(z)$ gives an equally valid $\omega^{\mathrm{IK}}(z)$, yielding the same value of Ω when integrated over z.

5.3 Capillary Waves

The Onsager theory *neglects* capillary wave fluctuations of the interface; these are also suppressed in computer simulations with insufficient transverse box dimensions. However, they may be detected in much larger-scale simulations [41]. Suppose that we split the simulation box, of overall dimensions $L_{\perp}\times L_{\perp}\times L_z$ into blocks (columns) of transverse dimension $\ell_{\perp}\times\ell_{\perp}$ and length L_z; for each block we measure the deviation $h(x,y)$ of the interfacial position from its average. The capillary wave Hamiltonian is given by the quadratic form:

$$\mathcal{H}_{\mathrm{CW}} = \frac{\gamma}{2}\int\int \mathrm{d}x\mathrm{d}y \left[\left(\frac{\partial h}{\partial x}\right)^2 + \left(\frac{\partial h}{\partial y}\right)^2\right] = \frac{\gamma}{2}\sum_{\boldsymbol{q}_\perp} q_\perp^2 |h(\boldsymbol{q}_\perp)|^2$$

with transverse wave-vector $\boldsymbol{q}_\perp = (q_x, q_y)$. Hence $\left\langle |h(\boldsymbol{q}_\perp)|^2 \right\rangle = k_{\mathrm{B}}T/\gamma q_\perp^2$ and the local mean-squared deviation of the interface may be obtained:

$$s^2 = \langle h^2(x,y)\rangle = \frac{1}{4\pi^2}\int \mathrm{d}\boldsymbol{q}_\perp \langle |h(\boldsymbol{q}_\perp)|^2\rangle = \frac{k_{\mathrm{B}}T}{2\pi\gamma}\ln\left(\frac{q_{\mathrm{max}}}{q_{\mathrm{min}}}\right)$$

with $q_{\mathrm{min}} = 2\pi/L_\perp$, $q_{\mathrm{max}} = 2\pi/\ell_\perp$. We expect a Gaussian distribution

$$P(h) = (2\pi s^2)^{-1/2}\exp(-h^2/2s^2)\,.$$

Moreover the intrinsic thickness of the interface, t_0, will be broadened, within a block of transverse dimension $\ell_\perp$, by capillary waves as follows

$$t^2 = t_0^2 + \frac{\pi}{2}\hat{s}^2 = t_0^2 + \frac{k_{\mathrm{B}}T}{4\gamma}\ln\left(\frac{\hat{q}_{\mathrm{max}}}{\hat{q}_{\mathrm{min}}}\right)$$

where $\hat{q}_{\mathrm{min}} = 2\pi/\ell_\perp$ and $\hat{q}_{\mathrm{max}} \sim 2\pi/a_0$, $a_0 =$ a microscopic length.

We studied a system of $N = 115200$ soft ellipsoids with $A/B = 15$, in a periodic box with dimensions $L_x = L_y = L_\perp \approx 150B$ and $L_z \approx 300B$. A nematic-isotropic film system was prepared and allowed to stabilize with the director assuming the preferred planar orientation: 1.2×10^6 MD steps were allowed for equilibration, and $\sim 2.0 \times 10^6$ MD steps to collect data. The local interfacial position distribution $P(h)$ is shown in Fig. 9: it is fitted well by the Gaussian distribution. The squared interfacial thickness, t^2, is plotted as a function of block size $\ell_\perp$, in Fig. 9. Particularly at large $\ell_\perp$, the dependence of t^2 on $\ell_\perp$ is logarithmic, which is consistent with the capillary wave picture. The interfacial tension estimated from this logarithmic dependence is in reasonable agreement with the values obtained from pressure tensor measurements.

We also study the difference in the capillary wave spectrum as a function of $\boldsymbol{q}_\perp$ with $\boldsymbol{q}_\perp$ in the plane of the interface, parallel and perpendicular to the director. Typical results are shown in Fig. 10. The low-$q_\perp$ behaviour is roughly consistent with the values of γ deduced from the interface width, and from the earlier pressure tensor calculation. The capillary wave spectrum is anisotropic at higher values of $q_\perp$. Fluctuations parallel to the director are less easy than those perpendicular to the director. The curvature in these figures indicates a *negative* bending rigidity.

6 Disclination in Cylindrical Pores

Recently, we have investigated the structure of disclination lines in a liquid crystal by confining a nematic phase of hard ellipsoids, once more with $A/B = 15$, in

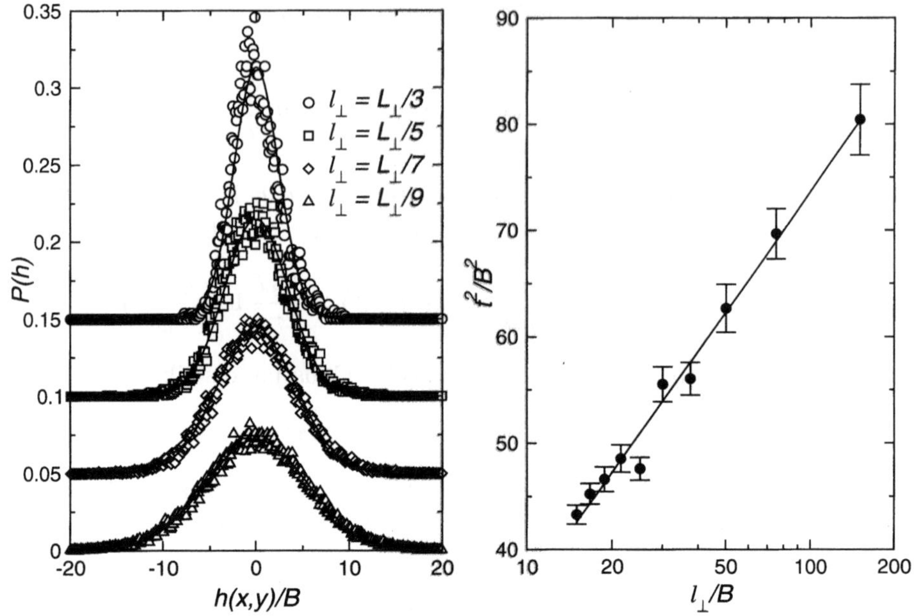

Fig. 9. Left: Probability distribution for interface height $P(h)$ averaged over blocks of various transverse dimensions (indicated). Curves are displaced vertically for clarity. **Right:** Mean squared interfacial thickness t^2 as a function of transverse dimensions of blocks (log scale). After [41]

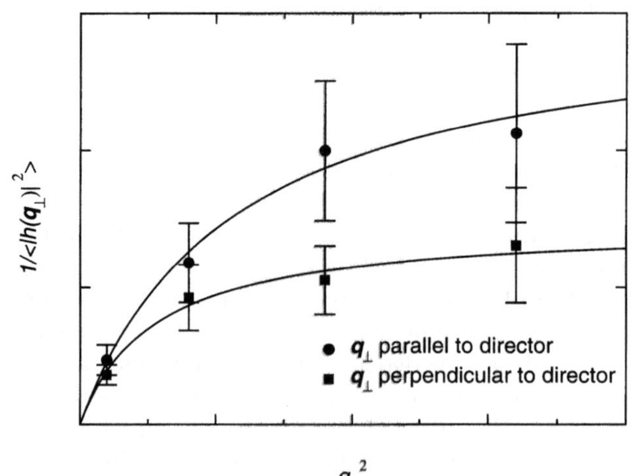

Fig. 10. Capillary wave spectrum as function of wavenumber, where the wave vector is directed respectively parallel and perpendicular to the director in the plane of the interface. The lines are to guide the eye. After [41]

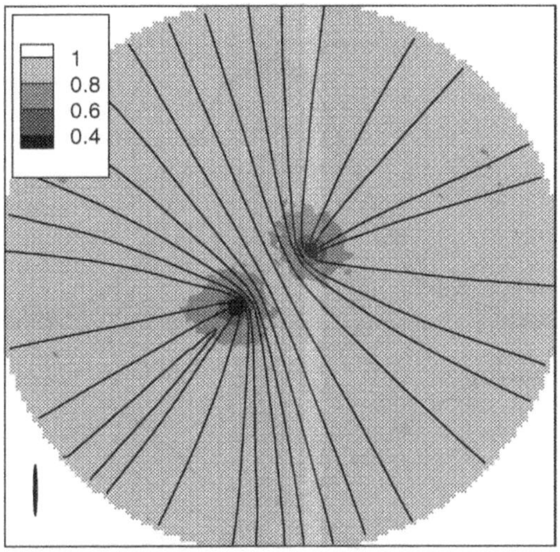

Fig. 11. Order parameter contour map and director field streamlines for pore radius $R/A = 5.33$. An indication of molecular size appears in the bottom left corner. For more details see [42].

a cylindrical pore, of radius R and height H [42]. With the aim of isolating finite size effects, four different cylinder radii were studied: $R/A = 2.08, 2.67, 4.00, 5.33$; in all cases the cylinder height was fixed at $H/A \approx 2.7$. This meant that the system size varied from $N = 3500$ to $N = 22000$. Homeotropic anchoring applied at the wall, as before. In these boundaries, a disclination line of strength +1 forms along the cylinder axis. The system is very similar to one originally studied several years ago [43]. Two of the most important cylinder-pore director structures are cylindrically symmetric: the *planar radial* defect, in which the director $\boldsymbol{n}(\boldsymbol{r})$ points along the $\boldsymbol{e}_\rho$ direction everywhere, and the *escaped radial* defect, in which the director is radial far from the cylinder axis, but bends over in such a way as to point in the $\boldsymbol{e}_z$ direction close to the axis. Examination of director field maps shows that the situation is not so simple. For example, for the $R/A = 5.33$ case, Fig. 11 reveals a pair of +1/2 defects close to the centre, each having a high degree of biaxiality. Notwithstanding this, the structure far from the core is indeed cylindrically symmetric, and since the defect separation is microscopic, it is reasonable to suppose that rotation of the structure about the axis should occur on a not-too-long timescale (but still, longer than the simulation runs). Accordingly, cylindrically averaged simulation averages may correspond to measurable properties of the system. Setting up a coordinate system with axes $\boldsymbol{e}_z$ (parallel to the cylinder axis), $\boldsymbol{e}_\rho$ (radially outward) and $\boldsymbol{e}_\theta$ (tangential, perpendicular to the other two), provides a natural frame of reference for the order tensor Q.

With this in mind, we show in Fig. 12 the time evolution (in Monte Carlo 'time' units of attempted moves per molecule, or sweeps) of the eigenvalues of the cylindrically averaged order tensor, for systems started from two different initial configurations. The perfectly ordered starting configuration has molecules

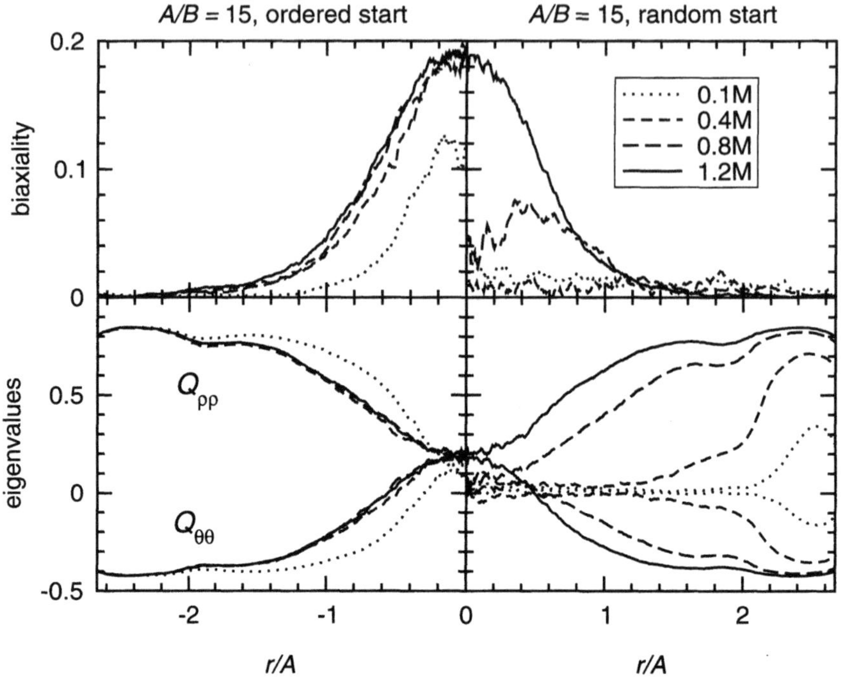

Fig. 12. Evolution of order tensor eigenvalues $Q_{\rho\rho}$ and $Q_{\theta\theta}$ (we omit Q_{zz} for clarity) for the cylinder pore system with $R/A = 2.67$, from a perfectly ordered planar radial starting configuration (left) and from a random, disordered starting configuration (right). Time is measured in millions of Monte Carlo sweeps. The development of the defect core is seen in the evolution of the biaxiality profiles. After [42]

placed randomly in space, avoiding overlaps, with molecular orientations fixed in the radial direction. This system evolves quite rapidly towards the planar radial defect structure by relaxation of the central core region. The disordered starting configuration has both positions and orientations of molecules chosen randomly, again avoiding overlaps. This evolves towards the same planar radial structure, but more slowly: the nematic phase grows inwards from the boundary. The general behaviour of Q-tensor eigenvalues can be easily understood in terms of the order map of Fig. 11. For these systems, with relatively small pore radius, the escaped radial configuration is not stable. We have confirmed this by starting from an escaped configuration, generated by applying a uniform aligning field in the $\boldsymbol{e}_z$ direction: following removal of the field, the system evolves rapidly to a planar director field.

Several theories are capable of reproducing the observed core structure. An example, based on the work of Schopohl and Sluckin [44,45], is shown in Fig. 13. This uses the Landau - de Gennes free energy, which is minimized as a function of order tensor variations.

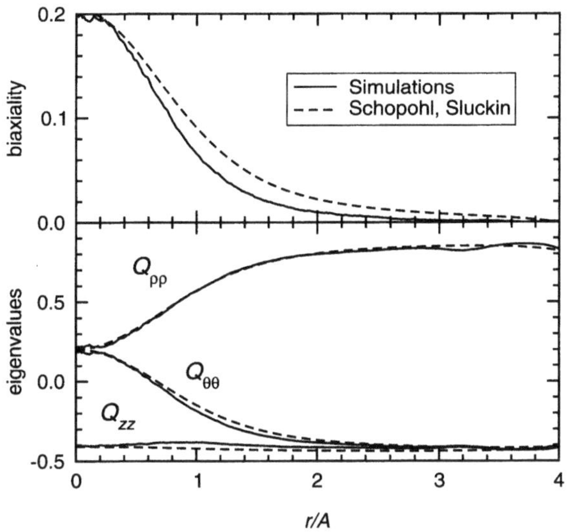

Fig. 13. Order tensor eigenvalues and biaxiality for cylinder defect simulation compared with predictions of the Landau-de Gennes theory of the kind proposed by Schopohl and Sluckin [44,45]. After [42]

7 Defects near a Colloid Particle

Colloidal particles or droplets suspended in a liquid crystal are of special interest [46]. Elastic deformations of the director field around the colloidal particles produce additional long-range forces between them [47–49]. These interactions can be of dipolar or quadrupolar type depending on the symmetry of the director field around the particles [50], and this in turn is extremely sensitive to the boundary conditions on the particle surface and the size of the particles [51]. The result may be supermolecular structures [46,52], cellular structures [53,54], and even a soft solid [55] can be observed. Colloidal dispersions in liquid crystals also have a wide variety of potential applications [56].

A key feature here is the presence of *topological defects* [50,57]. Strong homeotropic anchoring of the director, normal to the particle surface, induces a radial hedgehog defect with topological charge +1. If the director field is uniform far from the particle, i.e. the total charge of the whole system is zero, topological considerations imply that an additional defect must be created in the medium to compensate the radial hedgehog.

There are several types of defect which can arise in this case. Two are illustrated in Fig. 14. The first is a point defect: a hyperbolic hedgehog with a topological charge −1, called a dipolar or *satellite* defect. The second is a quadrupolar or *Saturn-ring* defect, i.e. a −1/2 strength disclination ring that encircles the spherical particle. Theoretical and numerical work based on elastic theory [58,57] suggest that the dipole configuration is stable for the micron-sized droplets which are usually realised experimentally; the Saturn-ring configuration

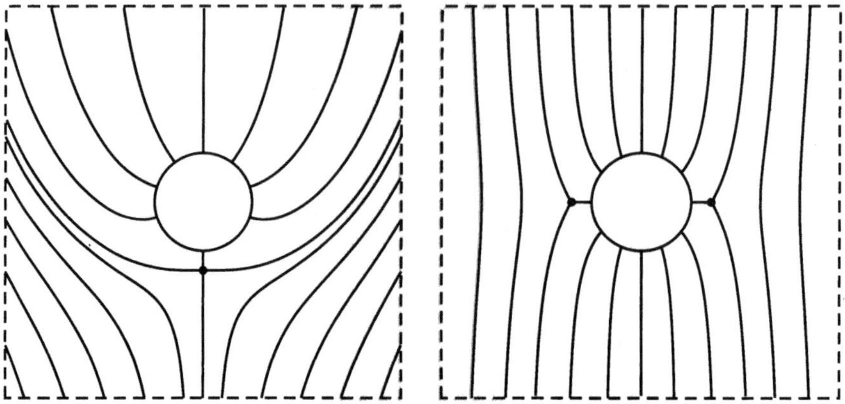

Fig. 14. Sketches of director field streamlines around a colloid particle, showing the satellite defect and Saturn ring defect

should appear if the droplet size is reduced or an external field is applied [59,60], and, when present, it is always predicted to be most stable in the equatorial plane normal to the director. The director distortion around the ring defect vanishes with distance in a manner consistent with the *quadrupolar* nature of the defect: far from the particle, at distance r in a direction θ relative to the director, the director deviation angle β has asymptotic behaviour $\beta \sim (R/r)^3 \sin 2\theta$ where R is the particle radius. The director distortion around the satellite defect extends much further than that of the ring defect. This is in accord with its symmetry: far from the particle, the director angle vanishes as $\beta \sim (R/r)^2 \sin\theta$, i.e. it is like a dipolar term in a multipole expansion. To minimize elastic interactions between images in computer simulations, rather larger periodic boxes are required in the satellite defect case.

Earlier investigations have simulated the Saturn ring type of defect for this kind of system [61], using 2048 Gay-Berne ellipsoids. We have recently looked at larger systems, investigating the structure for a range of colloid radii R, using soft ellipsoids with elongation $A/B = 3$ for the liquid crystalline solvent. The interaction of each molecule with the droplet was given by a shifted Lennard-Jones repulsion potential between the centers, rather similar to a softened version of the wall potential of Sect. 4. This results in homeotropic anchoring of the liquid crystal molecules, normal to the particle surface. The systems consisted of 8,000-1,000,000 mesogens. Full details appear in [62].

7.1 Saturn Ring Defect

For all studied radii in the range $R/B = 3$–15, the ring defect is relatively easy to prepare: it appears immediately after equilibration of the nematic phase of the system starting from the isotropic state, or on enlarging the droplet from a

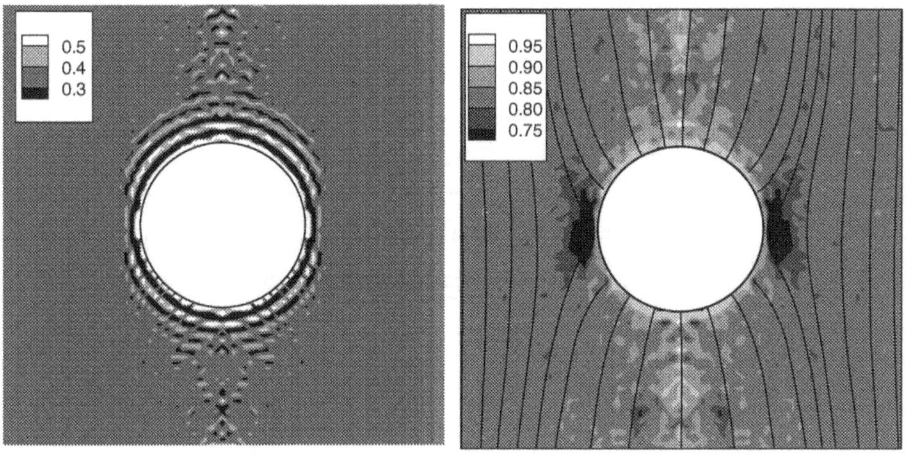

Fig. 15. Density map (left) and order parameter map with director streamlines (right) in the spatial region around the colloid particle, for the ring defect. Droplet radius $R/B = 15$. After [62]

small initial radius in the nematic solvent. It seems likely, therefore, that this type of the defect is energetically more favourable for the chosen droplet sizes.

Typical maps of density and order parameter are shown in Fig. 15. Strong density oscillations can be seen *except* in the region of the defect ring, which is identifiable by greatly reduced orientational ordering. Superimposed on the order parameter map are streamlines of the director field (note that these are not reliably defined in the defect regions themselves, where the order parameter is very low). More detailed information may be obtained by calculating profiles in radial slices. Typical density profiles for the ring defect with $R/B = 15$ are shown in Fig. 16 for directions $\theta = 0, \pi$ relative to the director (not intersecting the defect) and $\theta = \pi/2$ (crossing the disclination ring). The profiles which avoid the disclination have an oscillating structure near the particle surface which is typical for a *liquid-wall* interface. The profiles which cross the disclination ring do not have oscillations. The difference may be due to the partial melting of the liquid crystal in the disclination core region. This melting damps the influence of the droplet surface on the interface region.

The order parameter and biaxiality profiles for $\theta = \pi/2$ are shown in Fig. 16. The shape of these profiles in general reflects the typical structure of the core: the centre of the core has lower order than the bulk and the core region extends over a few molecular lengths. Both order parameter and biaxiality profiles indicate that the structure of the core is not simple.

Using the minima of the order profiles, we extracted the distance from the core of the disclination ring to the particle surface. The dependence of this distance on the particle radius R is shown in Fig. 17. It is interesting to compare with the phenomenological theory predicting linear dependence of the ring defect radius on the droplet radius: $r_{\text{ring}} \approx 1.25R$ from minimization of the elastic free energy using a trial function [51], $r_{\text{ring}} \approx 1.13R$ using a simulated annealing

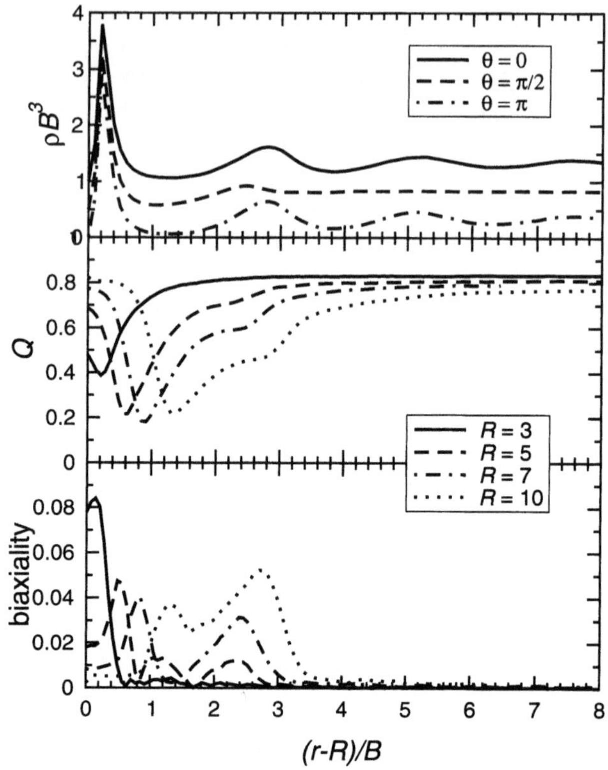

Fig. 16. Above: Density profiles for the ring defect. Droplet radius $R/B = 15$, $N = 64000$ particles. The following directions are shown: $\theta = 0, \pi$ (avoiding the defect) and $\theta = \pi/2$ (crossing the disclination ring). Successive curves are offset vertically by 0.5 for clarity. Below: Order parameter and biaxiality profiles for the ring defect along the direction $\theta = \pi/2$. Different curves correspond to the different droplet radii. After [62]

method [58], or $r_{\text{ring}} \approx 1.26R$ [57]. Our MD simulation results give $r_{\text{ring}} - R = -0.33B + (0.164 \pm 0.004)R$ which is in good agreement with the phenomenological theory, especially if one bears in mind the complex structure of the defect core.

7.2 Satellite Defect

To study the satellite defect , one needs very large systems. We used one million particles and droplet radius $R = 15B$. Below this radius, an initial configuration corresponding to the satellite defect structure would transform into the Saturn ring. (For $R = 10B$, this transformation occurred slowly enough to follow; see below).

Typical maps of density and order parameter are shown in Fig. 18. The density profiles for the satellite defect are shown in Fig. 19. One can see that the density profile at $\theta = \pi$ (across the defect core) has less prominent oscillating structure than the other two: this is again presumably due to the partial disor-

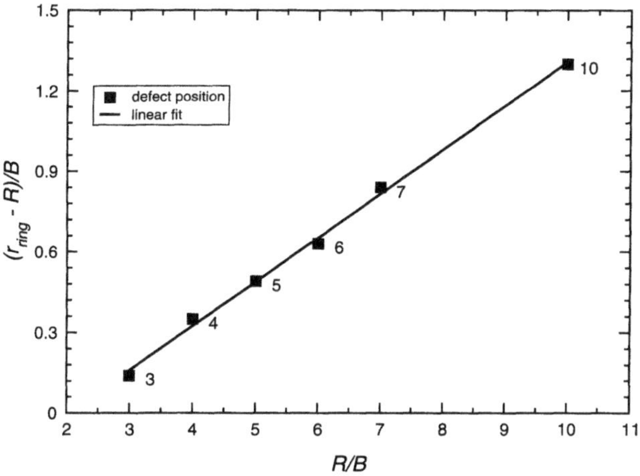

Fig. 17. Distance of the core region of the ring defect from the droplet center, as a function of reduced droplet radius R/B. We also show the linear fit to the simulation results, $r_{\text{ring}} = 1.164R - 0.33B$. After [62]

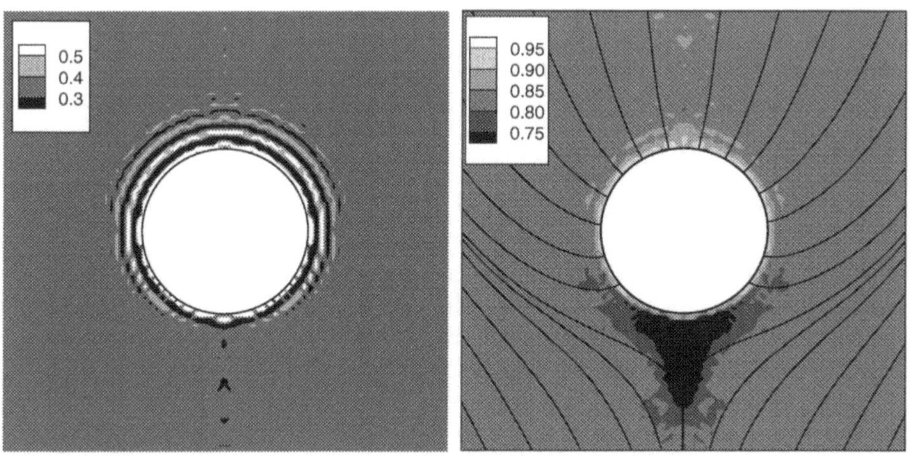

Fig. 18. Density maps and order parameter maps with director streamlines for the satellite defect. Droplet radius $R/B = 15$. After [62]

dering of the mesophase in the core region. The order parameter and biaxiality profiles in the $\theta = \pi$ direction are shown in Fig. 19. The defect is rather broad, and difficult to locate from these graphs. However, examination of the director field allows one to place the centre at a distance $r_{\text{sat}} \approx 1.4R$. The value predicted by the elastic theory is about $1.22R$ for the simulated annealing calculations [58]; $1.17R$ [46] or $1.46R$ [63] for the free energy minimization using a trial function.

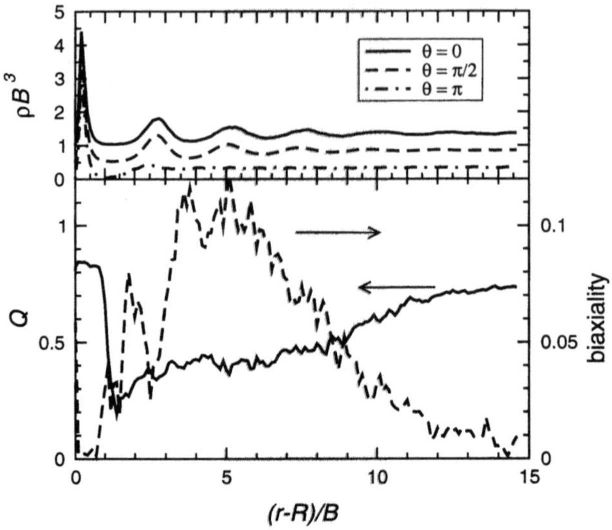

Fig. 19. Above: Density profiles for the satellite defect. Droplet radius $R/B = 15$. The following directions are shown: $\theta = 0, \pi/2$ (avoiding the defect) and $\theta = \pi$ (crossing the defect). Successive curves are offset vertically by 0.5 for clarity. Below: Order parameter and biaxiality profiles for the satellite defect along the direction $\theta = \pi$. After [62]

7.3 Off-Center Ring

Simulation results show that both the satellite and ring defects are at least metastable for $R/B = 15$: once the particular defect is realized in the system, it is stable over the timescale that is accessible to our simulations. However, the satellite defect is not stable for the smaller droplets. Indeed, we observed a rapid transition of the satellite defect to the ring defect for $R/B < 10$. Equilibrating the initial configuration with the satellite defect in the cell with droplet radius $R/B = 10$, we observed that it evolves into an *off-centered* ring defect. The ring then moved slowly, evolving towards an *equatorial* Saturn ring configuration of the usual kind. Doing long runs (up to a million timesteps) we conclude that the off-centred configuration is an intermediate state between the satellite and the Saturn ring defect. The z-coordinate of the ring, relative to the droplet center, as a function of the number of the timesteps, is shown in Fig. 20. The evolution dynamics is quite slow. Note that elastic theory also predicts the off-center ring configuration to be unstable [57], with the transition from the dipole configuration to the Saturn ring configuration occurring via this intermediate state.

7.4 Non-spherical Colloidal Particle

For a rod-like elongated particle with length L and transverse size $D \ll L$, and both L and D much greater than the dimensions of the molecules of the liquid crystal, one can have a disclination line of strength -1, a pair of disclination

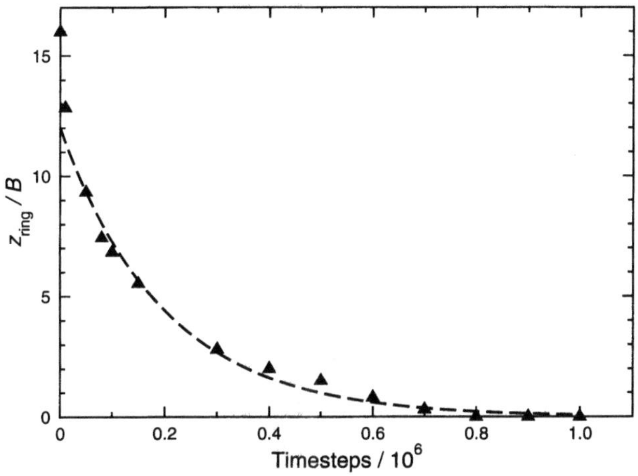

Fig. 20. Distance of centre of ring from centre of colloid, as a function of time, for $R/B = 10$. After [62]

lines of strength $-\frac{1}{2}$, as well as the 'escaped radial' structure, in which the director bends over to become perpendicular to the particle surface [64]. However, from the energetic point of view, the situation is different from the case of the spherical particle. For the elongated particle, both defects are disclination *lines*. The elastic energy per unit length associated with a disclination of strength m is $\pi K m^2 \ln(R/r_0)$, where R is the size of the sample and r_0 is a lower cutoff radius (the core size) [65]. This means that the free energy of a pair of $-\frac{1}{2}$ disclinations is always smaller than that of a single -1 disclination. Therefore, one can expect that the pair of $-\frac{1}{2}$ disclinations will always be a stable configuration, although, in principle, the -1 defect can still form a metastable state.

Phenomenological (continuum) theory [64] indicates that, depending on the type and strength of anchoring, the equilibrium orientation of the particle $\boldsymbol{s}$ may be either parallel or perpendicular to the liquid crystal director $\boldsymbol{n}$, depending on the ratio of the particle radius to the extrapolation length of the nematic liquid crystal. Qualitatively, the free energy should vary as

$$\mathcal{F} = \mathcal{F}_\perp + \left(\mathcal{F}_\parallel - \mathcal{F}_\perp\right) \cos^2\theta \, , \tag{2}$$

where $\cos\theta = \boldsymbol{n} \cdot \boldsymbol{s}$ and $\mathcal{F}_\parallel$ and $\mathcal{F}_\perp$ are constants. This gives a $\sin 2\theta$ dependence for the torque and predicts that the director response has a maximum at $\theta = \pi/4$ and is absent at $\theta = 0, \pi/2$. However, it is clear that this description is oversimplified: the defect structure changes while the particle rotates and the nematic ordering evolves in a complicated way.

Molecular dynamics simulations of $N = 64000$ repulsive, soft, ellipsoids of elongation 3 were carried out: the colloid particle was created in the nematic phase by slow expansion from nothing. For all studied diameters of the rod ($D/B = 5$–20) the -1 strength disclination appears immediately after expanding the colloid particle in the nematic state. However, during the equilibration, it

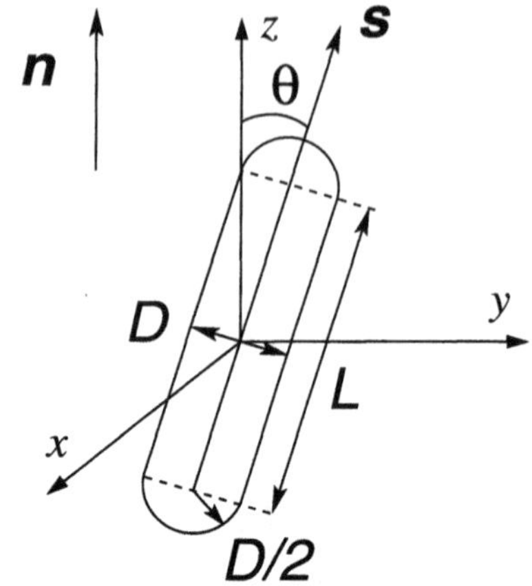

Fig. 21. Studied geometry: a spherocylinder of length L and diameter D is immersed in a liquid crystal host of Gay-Berne particles. The symmetry axis of the spherocylinder $\boldsymbol{s}$ is tilted with respect to the director $\boldsymbol{n}$, which defines the z axis.

splits into two $-\frac{1}{2}$ disclination lines which then move slowly towards the equatorial plane. We were not able to observe the 'escaped radial' configuration, probably due to the small size of the colloidal particle.

In the initial study, the colloidal rod particle and the pair of defect lines are perpendicular to the director: in this geometry the situation is like a two-dimensional version of the spherical colloid particle discussed earlier. Typical director streamlines together with the order parameter map are shown in Fig. 22. In the direction of the disclinations the density modulation, typical for a nematic-wall interface, vanishes due to partial melting of the liquid crystal in the disclination core region. This melting damps the influence of the droplet surface on the interface region. The two $-\frac{1}{2}$ disclinations are located very close to the droplet surface and the director distortion vanishes very quickly in the liquid crystal bulk. The core region extends over a few molecular lengths.

To emphasize the complex structure of the defect core we plot the eigenvalues Q_1, Q_2, Q_3, obtained by diagonalizing the local order tensor in Fig. 22. The nematic phase is uniaxial far from the core and biaxial in the core region, with variation of the biaxiality across the core and some complicated structure in the core region, due to the liquid crystal density oscillation near the particle surface.

To measure the torque on a rod of finite length, we performed MD simulations in a box with periodic boundary conditions, applying a global constraint for the director along the z axis. An independent measurement was performed in slab geometry, with the walls provided strong homeotropic (along the z axis) anchoring of the director, and with the rod located in the centre of the box. The rod diameter was $D = 5B$, the rod length $L = 10B$. The torque on the rod, as a function of tilt angle, is presented in Fig. 23. A positive torque tends to align the rod perpendicular to the director. The results presented indicate that the dependence of the torque on the rod tilt angle is far from the $\sin 2\theta$, proposed in [64]; indeed, the torque is not equal to zero for $\theta = 0°$, i.e. there is some

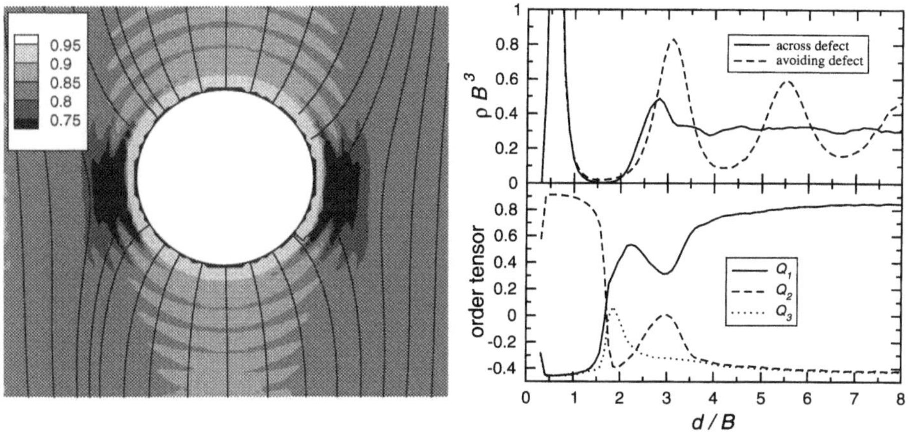

Fig. 22. MD simulation results for rod perpendicular to director: director streamlines of the xz cross section of the director field. Rod diameter $D = 20B$, rod length $L = \infty$. The shading represents the value of the order parameter. The director far from the particle is constrained along the z axis. A pair of $-\frac{1}{2}$ line defects forms parallel to the particle axis, perpendicular to the director. On the right we plot the density profile in slices across the defect and avoiding the defect (upper panel) and the three order tensor eigenvalues in a slice across the defect (lower panel). The density modulation near the particle affects the order parameter variation in the core region. For more details see [66].

symmetry breaking and the orientation of the rod along the director is not even metastable.

For better understanding, a slice in the yz plane is shown in Fig. 23, for different tilt angles of the rod. This shows that the director distribution around the rod is not axially symmetric, and hence the torque is nonzero, for $\theta = 0°$. As the rod rotates, the director field becomes less and less frustrated, and finally we have a stable orientation of the rod perpendicular to the director, $\theta = 90°$. In principle, a configuration with axial symmetry is also possible, when the rod is along the z axis. However, we were not able to observe it in our simulations even when disordered, isotropic configurations containing the colloidal particle were compressed to the ordered, nematic state. This method, in principle, gives the lowest free-energy configurations in an unbiased way. .

8 Conclusions

This chapter has attempted to cover several aspects of liquid crystal behaviour that are amenable to study by simulation and theory. It has concentrated on features of the structure which have length scales significantly larger than molecular sizes: defects, interfacial fluctuations, distortions due to suspended colloidal particles or droplets. Little mention has been made of many more aspects which are also of great interest: smectic (layered) phases, for instance, or reorientational

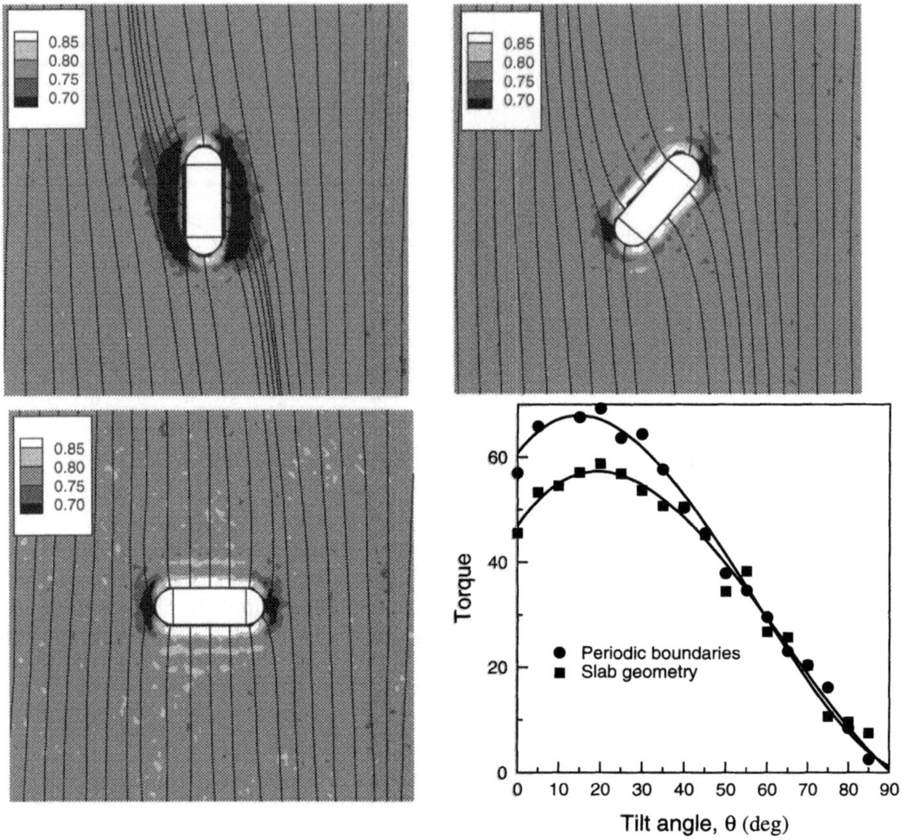

Fig. 23. Director streamlines and order parameter maps for different tilt angles. A side view along the x axis is shown (the rod is tilted in the yz plane). We also plot the torque on the rod vs. rod tilt angle for two systems: one with periodic boundaries and the other in slab geometry. The lines are to guide the eye. For more details see [66].

dynamics and fluid flow, which are important in understanding the operation of liquid crystal devices. These also provide rich fields of study, both by molecular simulation and other modelling techniques which may extend the accessible length and timescales towards the mesoscopic and macroscopic regimes. Many challenges remain in these areas. Finally, the predictive power of simulations remains limited: it is still quite a challenge to use atomistically correct potentials to predict, reliably, liquid crystalline phase behaviour and properties. Nonetheless, the progress that has been made using simple models is quite encouraging, and the steady increase in available computer power may bring these goals within reach in a few years.

Acknowledgments

It is a pleasure to acknowledge the work of my colleagues: M. S. Al-Barwani, D. Andrienko, M. Downton, G. Germano, A. J. McDonald, and F. Schmid, and also helpful conversations with R. Evans, H. Lange, R. van Roij and R. Roth. The author is grateful to K. Kremer (Max Planck Institute for Polymer Research, Mainz) and K. Binder (Department of Physics, Johannes Gutenberg University, Mainz) for their hospitality during a period of Study Leave, which was funded by a Fellowship from the Alexander von Humboldt foundation, and further supported by grants from the British Council ARC programme, and the Leverhulme Trust. This research was supported by grants from the UK Engineering and Physical Sciences Research Council.

References

1. J. G. Gay, B. J. Berne: J. Chem. Phys. **74**, 3316 (1981).
2. E. de Miguel, L. F. Rull, M. K. Chalam, K. E. Gubbins: Molec. Phys. **71**, 1223 (1990).
3. R. Berardi, A. P. J. Emerson, C. Zannoni: J. Chem. Soc. Faraday Trans. **89**, 4069 (1993).
4. E. de Miguel, E. Martín del Río, J. T. Brown, M. P. Allen: J. Chem. Phys. **105**, 4234 (1996).
5. J. T. Brown, M. P. Allen, E. Martín del Río, E. de Miguel: Phys. Rev. E **57**, 6685 (1998).
6. R. Berardi, C. Fava, C. Zannoni: Chem. Phys. Lett. **236**, 462 (1995).
7. P. J. Camp, M. P. Allen: J. Chem. Phys. **106**, 6681 (1997).
8. P. J. Camp, M. P. Allen, A. J. Masters: J. Chem. Phys. **111**, 9871 (1999).
9. J. S. van Duijneveldt, M. P. Allen: Molec. Phys. **92**, 855 (1997).
10. A. V. Lyulin, M. S. Al-Barwani, M. P. Allen, M. R. Wilson, I. Neelov, N. K. Allsopp: Macromolecules **31**, 4626 (1998).
11. J. S. van Duijneveldt, A. Gilvillegas, G. Jackson, M. P. Allen: J. Chem. Phys. **112**, 9092 (2000).
12. R. Berardi, M. Fehervari, C. Zannoni: Molec. Phys. **97**, 1173 (1999).
13. M. P. Allen, D. J. Tildesley: *Computer simulation of liquids* (Clarendon Press, Oxford, 1989), paperback ed., ISBN 0-19-855645-4, 385pp.
14. D. Frenkel, B. Smit: *Understanding molecular simulation : from algorithms to applications* (Academic Press, San Diego, 1996), ISBN 0–12–267370–0.
15. C. Oseen: Trans. Faraday Soc. **29**, 883 (1933).
16. F. C. Frank: Discuss. Faraday Soc. **25**, 19 (1958).
17. P. G. de Gennes, J. Prost: *The Physics of Liquid Crystals* (Clarendon Press, Oxford, 1995), second, paperback ed.
18. L. Onsager: Ann. N. Y. Acad. Sci. **51**, 627 (1949).
19. R. Evans: Adv. Phys. **28**, 143 (1979).
20. R. Evans, in *Liquids at Interfaces*, ed. by J. Charvolin, J. F. Joanny, J. Zinn-Justin (Elsevier Science Publishers B.V., Amsterdam, 1989), pp. 1–98, les Houches, Session XLVIII 1988.
21. R. Evans, in *Fundamentals of Inhomogeneous Fluids*, ed. by D. Henderson (Dekker, New York, 1992), chap. 3, pp. 85–175.

22. V. V. Ginzburg, M. A. Glaser, N. A. Clark: Liq. Cryst. **23**, 227 (1997).
23. W. H. Press, B. P. Flannery, S. A. Teukolsky, W. T. Vetterling: *Numerical Recipes in Fortran* (Cambridge University Press, Cambridge, 1992), 2nd ed.
24. M. P. Allen: Molec. Phys. **96**, 1391 (1999).
25. M. P. Allen, in *Advances in the Computer Simulations of Liquid Crystals*, ed. by C. Zannoni, P. Pasini (Kluwer Academic Publishers, Dordrecht, 2000a), vol. 545 of NATO *ASI Series C*, pp. 73–97, proceedings of the NATO Advanced Study Institute 'Advances in the Computer Simulations of Liquid Crystals', Erice, Italy, June 11-21, 1998. ISBN 0-7923-6099-0.
26. D. Andrienko, G. Germano, M. P. Allen: Phys. Rev. E **62**, 6688 (2000).
27. D. Andrienko, M. P. Allen: Phys. Rev. E (to appear).
28. A. Mertelj, M. Čopič: Phys. Rev. E **61**, 1622 (2000).
29. M. P. Allen: J. Chem. Phys. **112**, 5447 (2000b).
30. R. van Roij, M. Dijkstra, R. Evans: Europhys. Lett. **49**, 350 (2000a).
31. R. van Roij, M. Dijkstra, R. Evans: J. Chem. Phys. **113**, 7689 (2000b).
32. M. S. Al-Barwani, M. P. Allen: Phys. Rev. E **62**, 6706 (2000).
33. D. L. Koch, O. G. Harlen: Macromolecules **32**, 219 (1999).
34. P. Schofield, J. R. Henderson: Proc. Roy. Soc. Lond. A **379**, 231 (1982).
35. A. Harasima: Adv. Chem. Phys. **1**, 203 (1958).
36. J. H. Irving, J. G. Kirkwood: J. Chem. Phys. **18**, 817 (1950).
37. J. W. Perram, M. S. Wertheim, J. L. Lebowitz, G. O. Williams: Chem. Phys. Lett. **105**, 277 (1984).
38. J. W. Perram, M. S. Wertheim: J. Comput. Phys. **58**, 409 (1985).
39. A. J. McDonald, M. P. Allen, F. Schmid: Phys. Rev. E **63**, 010701(R)/1 (2000).
40. M. P. Allen: Chem. Phys. Lett. **331**, 513 (2000c).
41. N. Akino, F. Schmid, M. P. Allen: Phys. Rev. E **63**, 041706/1 (2001).
42. D. Andrienko, M. P. Allen: Phys. Rev. E **61**, 504 (2000).
43. S. D. Hudson, R. G. Larson: Phys. Rev. Lett. **70**, 2916 (1993).
44. N. Schopohl, T. J. Sluckin: Phys. Rev. Lett. **59**, 2582 (1987).
45. N. Schopohl: Phys. Rev. Lett. **60**, 755 (1988).
46. P. Poulin, H. Stark, T. C. Lubensky, D. A. Weitz: Science **275**, 1770 (1997).
47. B. I. Lev, P. M. Tomchuk: Phys. Rev. E **59**, 591 (1999).
48. A. Borštnik, H. Stark, S. Žumer: Phys. Rev. E **60**, 4210 (1999).
49. A. Borštnik, H. Stark, S. Žumer: Phys. Rev. E **61**, 2831 (2000).
50. T. C. Lubensky, D. Pettey, N. Currier, H. Stark: Phys. Rev. E **57**, 610 (1998).
51. O. V. Kuksenok, R. W. Ruhwandl, S. V. Shiyanovskii, E. M. Terentjev: Phys. Rev. E **54**, 5198 (1996).
52. P. Poulin, D. A. Weitz: Phys. Rev. E **57**, 626 (1998).
53. V. J. Anderson, E. M. Terentjev, S. P. Meeker, J. Crain, W. C. K. Poon: Euro. Phys. J. E **4**, 11 (2001).
54. V. J. Anderson, E. M. Terentjev: Euro. Phys. J. E **4**, 21 (2001).
55. S. P. Meeker, W. C. K. Poon, J. Crain, E. M. Terentjev: Phys. Rev. E **61**, R6083 (2000).
56. W. B. Russel, D. A. Saville, W. R. Schowalter: *Colloidal dispersions* (Cambridge University Press, Cambridge, 1989).
57. H. Stark: Euro. Phys. J. B **10**, 311 (1999).
58. R. W. Ruhwandl, E. M. Terentjev: Phys. Rev. E **56**, 5561 (1997).
59. Y. Gu, N. L. Abbott: Phys. Rev. Lett. **85**, 4719 (2000).
60. J. C. Loudet, P. Poulin: Phys. Rev. Lett. **87**, 165503 (2001).
61. J. L. Billeter, R. A. Pelcovits: Phys. Rev. E **62**, 711 (2000).

62. D. Andrienko, G. Germano, M. P. Allen: Phys. Rev. E **63**, 041701/1 (2001).
63. S. V. Shiyanovskii, O. V. Kuksenok: Mol. Cryst. Liq. Cryst. **321**, 489 (1998).
64. S. V. Burylov, Y. L. Raikher: Phys. Rev. E **50**, 358 (1994).
65. M. J. Stephen, J. P. Straley: Rev. Mod. Phys. **46**, 617 (1974).
66. D. Andrienko, M. P. Allen, G. Skacej, S. Zumer: Phys. Rev. E (to appear).

Disorder Diffuse Scattering of Crystals and Quasicrystals

Friedrich Frey

Ludwig-Maximilians-Universität, Institut für Kristallographie und Angewandte Mineralogie,
Theresienstr. 41, 80333 München

Abstract. Diffuse scattering is due to departures of any kind from a perfectly ordered periodic array of identical structural units in n-dimensional space. Crystals and quasicrystals refer to n = 3 and n > 3, respectively. Periodicity of aperiodically ordered quasicrystals is restored in higher dimensional (hyper-) space. The term "disorder" covers a wide range of structural fluctuations including displacive, chemical, or lattice (= domain) disorder. Phasonic type fluctuations in quasicrystals describe displacements of hyperatoms in the perpendicular subspace of the superspace. Bragg diffraction provides information about the space/time-average of a (quasi-) crystal, i.e. about properties of single atomic sites. Diffuse scattering contains information about how pairs of atoms or larger structural units behave. Examples of layer-disorder in crystals and disorder in decagonal quasicrystals are discussed in some detail.

1 Disorder

Structural disorder refers to partial loss of perfect long-range order (LRO) and/or presence of short-range order (SRO) in crystals. Perfect order means periodicity of structural units in conventional crystals or aperiodic order in quasicrystals. The terms "periodic" and "aperiodic" order may be reconciled if admitting a higher-dimensional description: then periodicity or, in other words a lattice is the common characteristic feature of all crystals including quasicrystals, and also incommensurate phases or composite structures. A loss of order may be highly anisotropic. For example, fiber-structures or channel-like structures exhibit LRO only along a unique direction and with only SRO perpendicular to it. SRO refers to limited correlations between structural units which are in simplest case single atoms (Fig. 1). The atomic arrangement in a SiO_2-glass has perfect SRO, but no more LRO (Fig. 2). On the other hand, stacks of block-co-polymers, or more simply, an array of carefully stacked sacks of potatoes, show significant LRO, but no SRO. Any combination of LRO and SRO is possible, and there is – in general – no continuous transition between SRO and LRO via disorder. The inherent basic aspect refers to the element of an **average** structure: the array of potato sacks shows average LRO if averaging over all sizes and positions of the potatoes within one sack. As introduced in [24] one may therefore distinguish between disorder of first and second kind if an average structural element with LRO either still exists or is gradually lost.

Replacement disorder, i.e *chemical, compositional, substitutional* disorder relates to structural units, in the simplest case atoms, which are replaced by an-

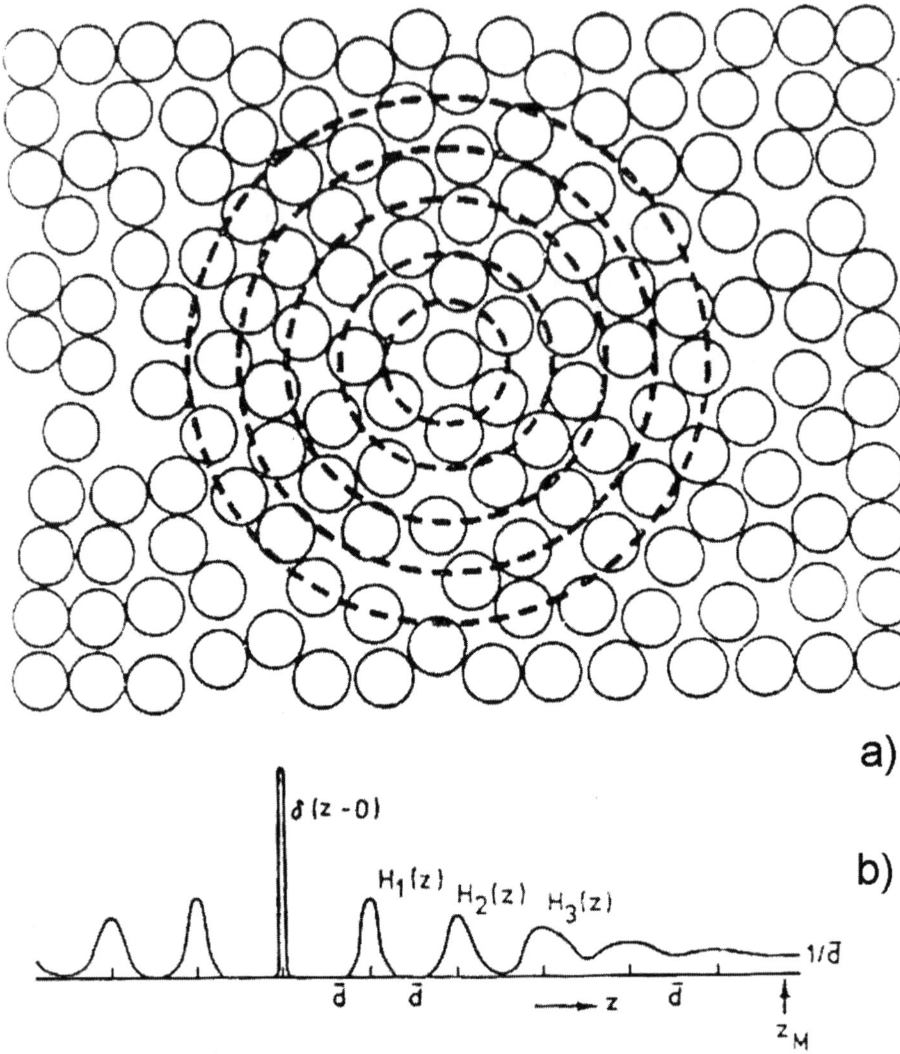

Fig. 1. Short-range order in a disturbed array of equal spheres (a); nearest neighbour distribution as indicated by a radial section outgoing from a reference atom (b)

other species (including vacancies). This type of disorder is commonly present in, for example, alloys or mixed crystals (Fig. 3).

Displacement or *positional* disorder covers a wide range of various phenomena: local distortions of atoms, straining, modulated structures, thermal vibrations (phonon-like disorder), paracrystalline order etc. Disorder elements specific to quasicrystals, so-called phasons, will be discussed separately (Sect. 5).

Orientational disorder relates to the orientation of extended anisotropic molecules or larger structural subunits where the centres of these units obey strict

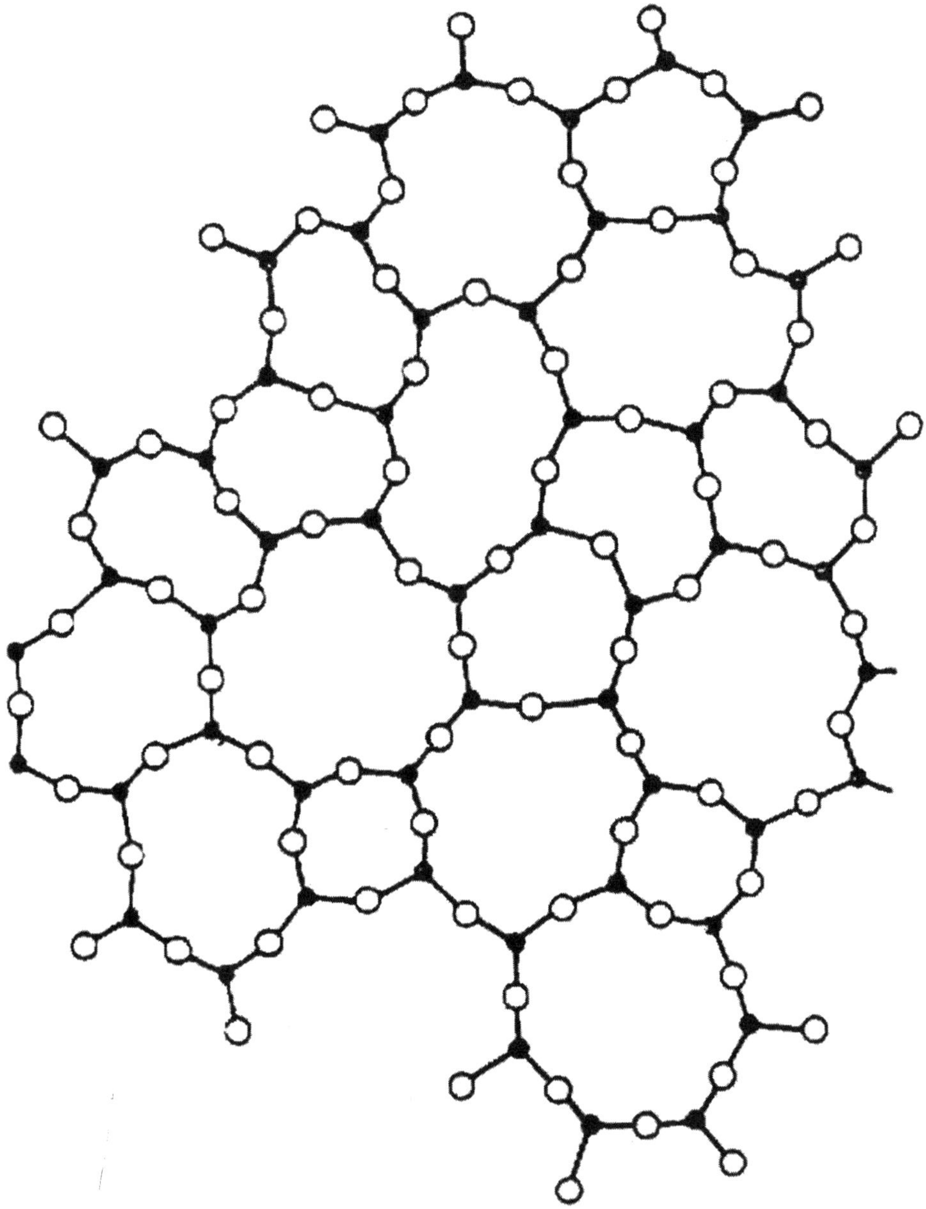

Fig. 2. Schematic picture of a SiO_2 glass network

periodicity. Prominent examples are liquid or plastic crystals, and fullerites, where the constituents, the C-"buckyballs", change their orientation as function of the temperature [25]. There is a wealth of order/disorder phenomena in the mesomorphic phases (liquid crystals).

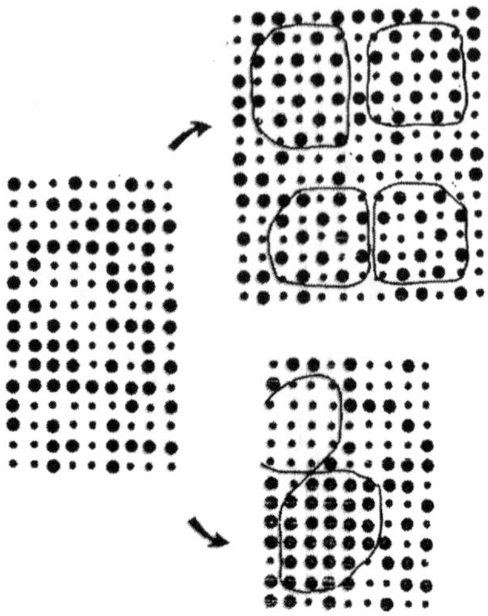

Fig. 3. Binary alloy with a tendency either to segregation (clustering) or to superordering in domains

Lattice disorder refers to regular displacements of parts of a structure by large fractions of a lattice vector. Parts of a structure may be sheets of atoms, twin variants, polytypic variants, or, in general, domains (an example is shown in Fig. 5). Domains may exhibit a tendency to periodic superordering with a limited correlation length of the superperiod. The related diffraction patterns show more or less diffuse satellite reflections. A further discussion of this aspect is beyond the scope of this contribution (see [12]). Another aspect of this type of disorder concerns the walls between different domains. Usually there are matching problems between them. As a consequence strains or other defects may occur which accommodate the mismatch and which are responsible for the degree of coherence between the domains (see Fig. 6). The topic of domain disorder will be further discussed in Sect. 3.

Another class of disorder phenomena relates to *composite structures* which consist of two (or even more) substructures. These substructures may differ, for example, by different (incommensurate) lattices, different symmetries, different dimensionalities, such as chains in tubes of a matrix structure, or completely different substructures including liquid-like atomic arrangements.

From a physical point of view the origin of a particular disorder problem may be different due to energetic or entropic reasons. Examples are: disordering as a consequence of frustrated coordination (chemical bonding), as a result or a precursor of phase transitions or phase transformations, as a consequence of strain release, as a consequence of the electronic band structure in relation the shape of the Fermi surface (Hume-Rothery mechanism in "electron-compounds", formation of domains in alloys), or as a consequence of coherent or incoherent intergrowth of different chemical constituents during the crystal growth process – in natural "unclean" conditions or in a chemical laboratory. Many disordered

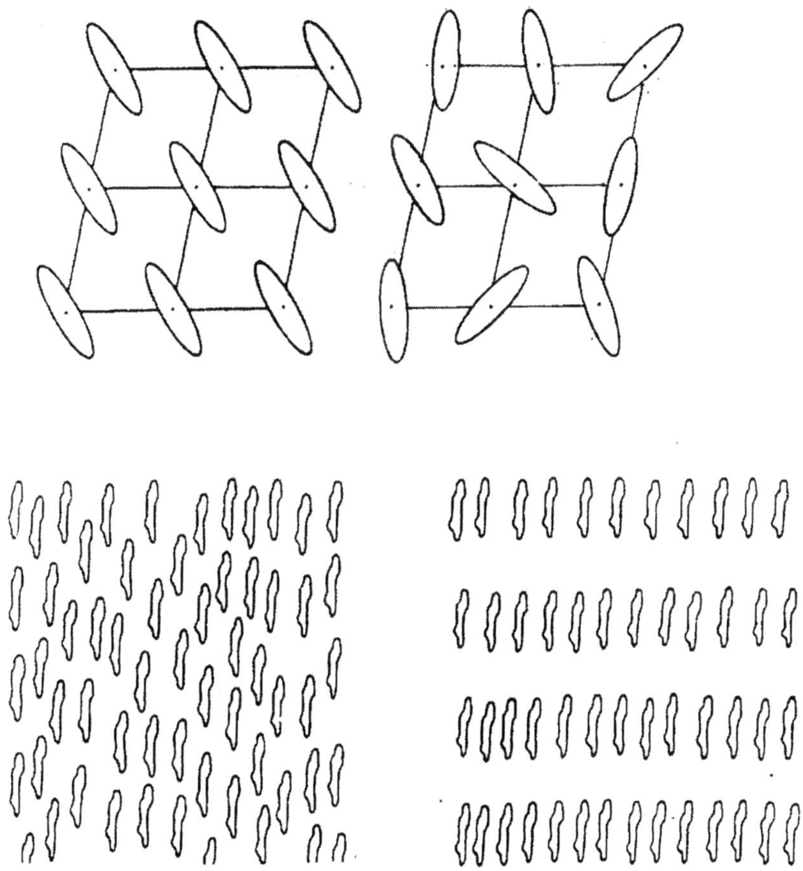

Fig. 4. Principle of orientational disorder; two examples of intermediate ordering in mesophases (liquid crystals)

structures are due to external field conditions, such as a shock wave or high-temperature/high pressure conditions. Thus the analysis of disorder problems gives an opportunity to learn about the prior history of the particular sample. On the other hand, the properties of disordered structures may be different, which offers the opportunity of tailoring material properties. These remarks only elucidate the background behind the subject of disorder, and will not be discussed further. This article is devoted to a powerful tool of analysing disorder, i.e. the diffuse diffraction method, irrespective of the origin of a particular phenomenon. The following text is organised in four main sections: (2) basic diffraction formulae of diffuse scattering, (3) specific problems of diffuse scattering by domain-disordered structures, (4) diffuse scattering by layer or lamellar type of disorder, and (5) scattering by disordered quasicrystals. There are several review articles about diffuse scattering in conventional crystals which can

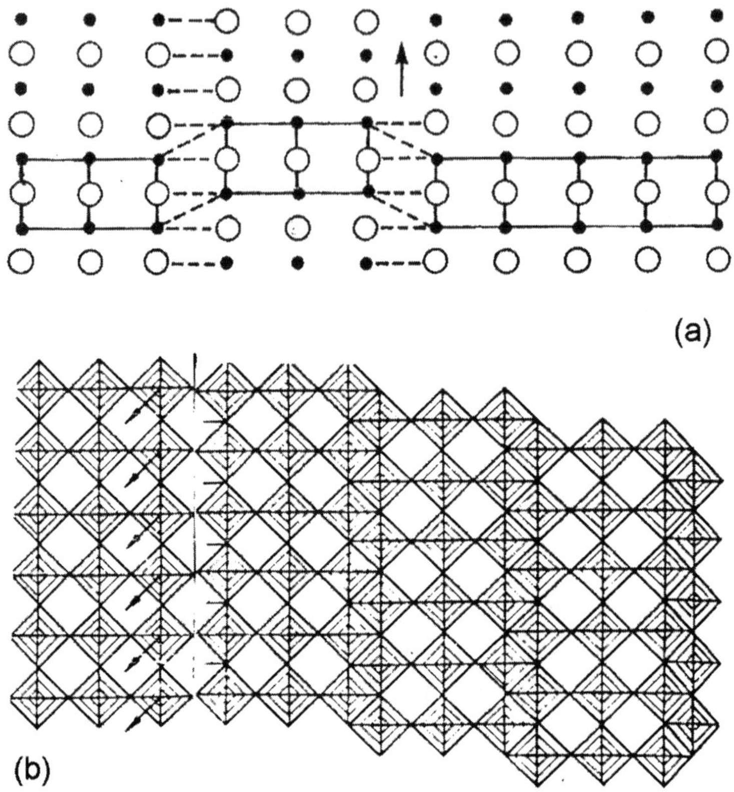

Fig. 5. Principle of lattice disorder (a); example of shear domains (b)

be recommended for further reading, for example [12,5,20,22,26] The last topic (5) is of current research interest, an introduction is given in [27].

2 Basic Diffraction Theory

There is no general theory which is well applicable to the various disorder problems. Generally, diffraction phenomena can be described in an analytical way by the use of Fourier transforms. Vectors in real and reciprocal (=Fourier) space are denoted by $\boldsymbol{r} = (x, y, z)$ and $\boldsymbol{Q} = (\xi, \eta, \zeta)$, respectively. The coordinates (x, y, z) and (ξ, η, ζ) refer to arbitrary values in units of basis vectors in respective space.

$$A(\boldsymbol{Q}) = \int a(\boldsymbol{r}) \exp\{2\pi i \boldsymbol{Q}\boldsymbol{r}\}\, d\boldsymbol{r} \quad \text{and} \quad a(\boldsymbol{r}) = \int A(\boldsymbol{Q}) \exp\{-2\pi i \boldsymbol{Q}\boldsymbol{r}\}\, d\boldsymbol{r} \quad (1)$$

briefly written as $a(\boldsymbol{r}) \leftrightarrow A(\boldsymbol{Q})$. We use the symbols $\rho(\boldsymbol{r})$ and $F(\boldsymbol{Q})$ for structure (of a unit cell) and corresponding structure factor, respectively.

2.1 Elementary Formulae

For the purpose of this article some useful laws of Fourier transformation are:

$$a(-\boldsymbol{r}) \leftrightarrow A(-\boldsymbol{Q}) \quad \text{which means, if } a(\boldsymbol{r}) = a(-\boldsymbol{r}) \quad \text{then} \quad A(\boldsymbol{Q}) = A(-\boldsymbol{Q})$$

$$a^{+}(\boldsymbol{r}) \leftrightarrow A^{+}(-\boldsymbol{Q}),\ a^{+}(-\boldsymbol{r}) \leftrightarrow A^{+}(\boldsymbol{Q})$$

where "+" denotes "complex conjugate" (c.c.).

Complex functions have to be considered if anomalous X-ray scatterers are present in a structure:

$$a(\boldsymbol{r}) + b(\boldsymbol{r}) \leftrightarrow A(\boldsymbol{Q}) + B(\boldsymbol{Q})$$

$$sa(\boldsymbol{r}) \leftrightarrow sA(\boldsymbol{Q}) \quad (\text{s} = \text{scalar quantity})$$

$$a(\boldsymbol{r} - \boldsymbol{r}_o) \leftrightarrow \exp\{2\pi i \boldsymbol{Q}\boldsymbol{r}_o\}A(\boldsymbol{Q}) \quad \text{and} \quad A(\boldsymbol{Q} - \boldsymbol{Q}_o) \leftrightarrow \exp\{-2\pi i \boldsymbol{Q}\boldsymbol{r}_o\}a(\boldsymbol{r}).$$

Translations in either space correspond to phase factors in the respective dual space. This gives rise to characteristic extinction rules in diffraction patterns

$$a(\boldsymbol{r}) \cdot b(\boldsymbol{r}) \leftrightarrow A(\boldsymbol{Q}) * B(\boldsymbol{Q})$$

$$a(\boldsymbol{r}) * b(\boldsymbol{r}) \leftrightarrow A(\boldsymbol{Q}) \cdot B(\boldsymbol{Q})$$

$$a(\boldsymbol{r}) * a^{+}(-\boldsymbol{r}) \leftrightarrow |A(\boldsymbol{Q})|^2 \quad (\text{Note: } |A(-\boldsymbol{Q})|^2 \neq |A(\boldsymbol{Q})|^2 \text{ if } a(\boldsymbol{r}) \text{ is complex}).$$

The symbol * denotes the operation of convolution:

$$a(\boldsymbol{r}) * b(\boldsymbol{r}) = \int a(\boldsymbol{r}') \cdot b(\boldsymbol{r} - \boldsymbol{r}')\, d\boldsymbol{r}';\ A(\boldsymbol{Q}) * B(\boldsymbol{Q}) = \int A(\boldsymbol{Q}')B(\boldsymbol{Q} - \boldsymbol{Q}')\, d\boldsymbol{Q}'$$

$$a(\boldsymbol{r}) * [b(\boldsymbol{r}) + c(\boldsymbol{r})] = a(\boldsymbol{r}) * b(\boldsymbol{r}) + a(\boldsymbol{r}) * c(\boldsymbol{r}).$$

The associative law of multiplication is not valid for mixed products:

$$[a(\boldsymbol{r}) * b(\boldsymbol{r})] \cdot c(\boldsymbol{r}) \neq a(\boldsymbol{r}) * [b(\boldsymbol{r}) \cdot c(\boldsymbol{r})].$$

Particularly important functions are:

1. Patterson function (or autocorrelation function) of a structure $\rho(\boldsymbol{r})$:

$$P(\boldsymbol{r}) = \rho(\boldsymbol{r}) * \rho^{+}(-\boldsymbol{r}) \quad (\text{Note: } P(\boldsymbol{r}) = P(-\boldsymbol{r}) \text{ holds only if } a(\boldsymbol{r}) \text{ is real !}),$$

$$P(\boldsymbol{r}) \leftrightarrow |A(\boldsymbol{Q})|^2 \sim I(\boldsymbol{Q}) \quad (= \text{"intensity" apart from other factors}).$$

2. δ-"functions" $\delta(\boldsymbol{r}), \delta(\boldsymbol{Q})$:
We use here its "relaxed definition" $\delta(\boldsymbol{r}) = 1$ if $\boldsymbol{r} = 0$ and zero elsewhere:

$$\delta(\boldsymbol{r}) \leftrightarrow 1 \quad (\text{const.}), \text{ and} \quad \delta(\boldsymbol{Q}) \leftrightarrow 1.$$

In consequence, sharp phenomena in real space correspond to broad phenomena in reciprocal (Fourier-) space and vice versa.

$$\delta(\boldsymbol{r} - \boldsymbol{r}_o) * \delta(\boldsymbol{r} - \boldsymbol{r}'_o) = \delta(\boldsymbol{r} - \boldsymbol{r}_o - \boldsymbol{r}'_o)$$

$$\delta(\boldsymbol{r} - \boldsymbol{r}_o) \leftrightarrow \exp\{2\pi i \boldsymbol{Q}\boldsymbol{r}_o\}.$$

3. Lattice functions:
   Finite lattice (basis $\boldsymbol{a}_1, \boldsymbol{a}_2, \boldsymbol{a}_3; \boldsymbol{t}_n = n_1\boldsymbol{a}_1 + n_2\boldsymbol{a}_2 + n_3\boldsymbol{a}_3 = (n_1, n_2, n_3)$)

$$\sum(\boldsymbol{r} - \boldsymbol{t}_n) \leftrightarrow \sum \exp\{2\pi i \boldsymbol{Q}\boldsymbol{t}_n\} = \sin(\pi N_1 h)/\sin(\pi h)\{..N_2..\}\{..N_3..\}$$

   where the summation over n_1, n_2, n_3 extends to N_1, N_2, N_3, respectively.
   Infinite lattice:

$$l(\boldsymbol{r}) = \sum \delta(\boldsymbol{r} - \boldsymbol{t}_n) \leftrightarrow L(\boldsymbol{Q}) = C\sum \delta(\boldsymbol{Q} - \boldsymbol{H}) \quad \text{(reciprocal lattice)}$$

   with $\boldsymbol{H}$ = reciprocal lattice vector; integer coordinates h,k,l; reciprocal basis vectors $\boldsymbol{a}_1^*, \boldsymbol{a}_2^*, \boldsymbol{a}_3^*$. The normalising factor C will be omitted in the following.

2.2 Average Structure, Structural Fluctuations

Crystal structures with small, e.g. thermally excited, fluctuations are most common. In such cases a separation into two parts, an average one and deviations from the average, is feasible.

Scattering density:

$$\rho(\mathbf{r}) = \langle \rho(\mathbf{r}, t) \rangle_{r,t} + \Delta\rho(\mathbf{r}, t) \qquad (\langle \Delta\rho \rangle = 0). \tag{2}$$

The average fluctuations may be time dependent or not, i.e. dynamic or static. Fourier transformation gives:

$$FT\{\rho\} \equiv F(\mathbf{Q}, \omega) = FT\{\langle \rho \rangle\} + FT\{\Delta\rho\} \equiv \langle F(\mathbf{Q}) \rangle + \Delta F(\mathbf{Q}, \omega) \tag{3}$$

where ω = energy.

Average structure:

$$\langle \rho \rangle = \langle \rho_c \rangle * l(\boldsymbol{r}); \quad \langle \rho \rangle = \frac{1}{N}\sum_n \rho_n, \tag{4}$$

$\langle \rho \rangle$ denotes the average over all N unit cells.

$$\langle A \rangle = \langle F \rangle \cdot L(\boldsymbol{Q}) \tag{5}$$

$$\langle F \rangle = \frac{1}{N}\sum_n F_n \exp\{i\boldsymbol{Q}\boldsymbol{t}_n\}, \tag{6}$$

$\langle F \rangle$ denotes the average over all individual structure factors.
Patterson function (time independent fluctuations):

$$\begin{aligned}\langle \rho(\mathbf{r}) * \rho(-\mathbf{r}) \rangle &= \langle (\langle \rho \rangle + \Delta\rho) * (\langle \rho \rangle + \Delta\rho) \rangle \\ &= \langle \rho \rangle * \langle \rho \rangle + \langle \Delta\rho * \Delta\rho \rangle\end{aligned}$$

$$\begin{aligned}\mathrm{I}_{total} &= \langle A \rangle \langle A^* \rangle + \langle \Delta A \Delta A^* \rangle \\ &= |\langle F \rangle|^2 L + \langle \sum_n \sum_m \Delta F_n \Delta F_m^* \exp\{i\mathbf{Q}(\mathbf{r_n} - \mathbf{r_m})\} \rangle \\ &= \mathrm{I}_{Bragg} + \mathrm{I}_{diffuse},\end{aligned} \tag{7}$$

$$I_{diff} = < \sum_n \sum_m \Delta F_n \Delta F_m^* \exp\{i\boldsymbol{Q}(\boldsymbol{r_n} - \boldsymbol{r_m})\}$$
$$= \sum_k (\sum_n \Delta F_n \Delta F_{n+k}^*) \exp\{i\boldsymbol{Q}(\boldsymbol{r_k})\} >, \tag{8}$$

$$I_{diff} = \sum_k \langle \Delta F_n \Delta F_{n+k}^* \rangle_n \exp\{i\mathbf{Q}(\mathbf{r_k})\} > \tag{9}$$

with: $\Delta F = \Delta f \exp\{i\mathbf{Q}(\mathbf{r} + \Delta\mathbf{r})\}$,

$\Delta f \rightarrow$ substitutional (replacement) disorder,
$\Delta\mathbf{r} \rightarrow$ displacement disorder.

From (9) it becomes clear that diffuse scattering is related to correlations between pairs of structural units (atoms) which are k cells apart from one another.

Examples:

- small *displacements* $\Delta\boldsymbol{r_n}$: $\exp\{i\boldsymbol{Q}\Delta\boldsymbol{r_n}\} \approx (1 + i\boldsymbol{Q}\Delta\boldsymbol{r_n} + ...)$,

$$I_D \approx N f_2 \sum_k \sum_n (\boldsymbol{Q}\Delta\boldsymbol{r_n})(\boldsymbol{Q}\Delta\boldsymbol{r_{n+k}}) \exp\{i\boldsymbol{Q}\Delta\boldsymbol{r_k}\} \tag{10}$$

  – (thermal) random vibrations

$$I_D \sim N f^2 \boldsymbol{Q^2} \langle \Delta\boldsymbol{r_n^2} \rangle_n [= (1 - \exp\{-2W\})] \tag{11}$$

  – periodic modulations

$$\Delta\boldsymbol{r_n} = \boldsymbol{\Delta} \cos(\boldsymbol{q}\boldsymbol{r_n});\ I_D = (\boldsymbol{Q}\boldsymbol{\Delta})^2 \sum_{\boldsymbol{Q}} \delta(\boldsymbol{Q} - \boldsymbol{Q} \pm m\boldsymbol{q}) \tag{12}$$

  (i.e. "satellite" scattering at positions $m\boldsymbol{q}, m = \pm 1, \pm 2, ...$).
- *substitutional* disorder (without atomic displacements):
  c_{ij} = concentrations of species i, j; p_n^{ij} = conditional probability to find species j at distance $\boldsymbol{n}$ from a site occupied by species i; $\boldsymbol{n}$ = lattice vector)
  – random:

$$I_D = N\langle \Delta F^2 \rangle = N - c \sum_i \sum_j c_i c_j f_i f_j^* [p_0^{ij}/c_j - 1] \tag{13}$$

  – random, binary

$$I_D = N_c c_A c_B (f_A - f_B)^2 \tag{14}$$

  Laue monotonic scattering.
  – short-range correlated

$$I_D (= I_{sro}) \sim \sum_n \sum_i \sum_j c_i c_j f_i f_j \alpha_n^{ij} \cos(\boldsymbol{Q}\boldsymbol{n}) \tag{15}$$

$$\tag{16}$$

  with: $\alpha_n^{ij} = 1 - (p_n^{ij}/c_j)$ "Warren-Cowley sro parameter".
  I_D is concentrated around/beneath Bragg- or superstructure reflections.

– sro-defect pair at distance $\boldsymbol{\Delta}: I_D \sim c_1 c_2 (f_1 - f_2)^2 (1 + \cos(\boldsymbol{Q}\boldsymbol{\Delta}))$
– periodic modulation: $\Delta f_n = f_o \cos(\boldsymbol{q}\boldsymbol{r_n})$
satellite pair: $I_D \sim f_o^2 \sum_Q \delta(\boldsymbol{Q} - \boldsymbol{H} \pm \boldsymbol{q})$.

- *short-range order + accompanying static displacements (atomic size effect):*

$$I_D \sim \sum_n \sum_i \sum_j N_n c_i c_j f_i f_j^* [p_n^{ij}/c_j \langle \exp\{i\boldsymbol{Q}\boldsymbol{u}_{\boldsymbol{n}}^{ij}\}\rangle$$
$$-\langle \exp\{i\boldsymbol{Q}\boldsymbol{u}^{ij}\}\rangle] \exp\{i\boldsymbol{Q}\boldsymbol{n}\}, \tag{17}$$

$\boldsymbol{u}^{ij}$ denotes the average over the displacements over all valid interatomic vectors. If only local deviations exist, the expression in square brackets vanishes at large distances. Assuming that $\boldsymbol{u}_{\boldsymbol{n}}^{ij}$ and $\boldsymbol{u}^{ij}$ are small, a series expansion of the exponentials allows for the separation:

$$I_D \approx I_{sro} + I_1 + I_2 \tag{18}$$

where I_1 and I_2 include the displacements (first and second order). For a detailed discussion see [26].

2.3 Disorder Without Average Structure (Disorder of 2nd Kind)

Formally, an average structure can always be derived. However, the separation becomes more or less meaningless if the average part of the disordered structure is (close to) zero. Then all reflections are dominated by diffuse contributions and a Bragg-like part cannot be separated. In that case the method outlined in the foregoing section is less helpful and different methods have to be applied. If there is a gradual loss of long-range order, the lattice function has to be replaced by a distribution.

As an example, the distribution of molecules ρ_{mol} along a string may be described by a distribution $d(z)$ of the 1st neighbour (paracrystal model, see [28]). Then the mean distance is $\bar{c} = \int z d(z) dz$ and the position of the mth neighbour is given by: $d(z) * d(z) * d(z) \ldots * d(z)$ (Fig. 6).

Intensity:	$I(l) \sim \lvert F_{mol}\rvert^2 \cdot \lvert D(l)\rvert^2$	
Model:	$d(z) = 1/\pi\eta^2 exp\{-z^2/\eta^2\}$	(gaussian)
m^{th}peak:	height $\sim \lvert D(m)\rvert^2 \sim 1/m^2$;	width $\sim m^2$.

3 Domain Disorder

3.1 Domains

Domain formation in a crystal describes a spatial arrangement of different regions which are related to one another in some regular way and which are separated

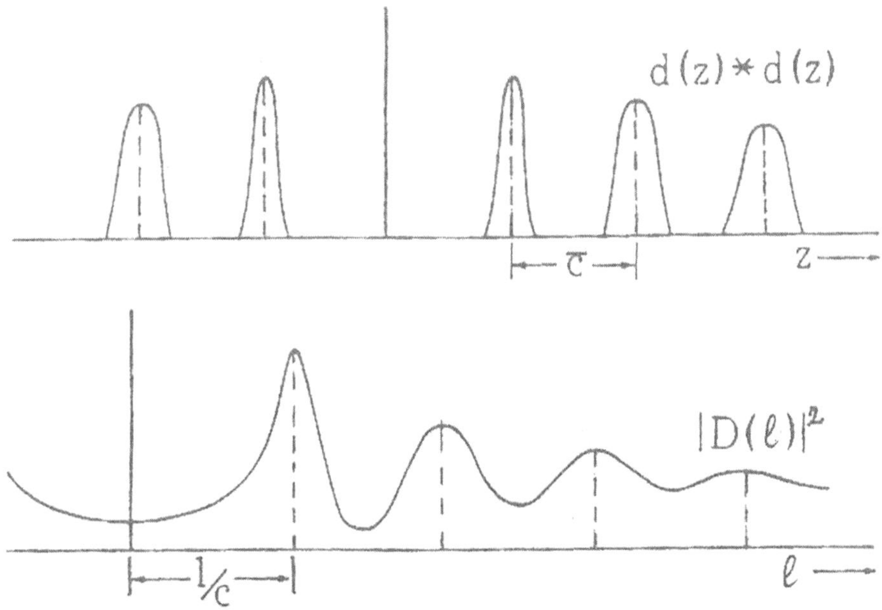

Fig. 6. Atomic distribution and corresponding interference function of a 1D paracrystal model

by walls. Structural domains may carry a physical property in different "configurations", such as the orientations of an electric dipole moment in ferroelectric domains. Whereas an electric dipole moment has a structural origin viz. the separation of the centre of gravities of negatively and positively charged atoms or molecules, magnetic domain structures refer to ordering of magnetic moments.

The array of these moments may exhibit a domain pattern irrespective of the underlying chemical structure. Structural domains may be chemically homogeneous, as in the case of twin domain structures, or chemically heterogeneous. This aspect can be visualised easily by considering a binary alloy consisting of two chemical species A, B. Depending on the interaction energy between A and B, dissimilar atoms tend to be neighbours or not. Either we have a tendency for A-B ordering in domains which are embedded in an otherwise disordered A/B matrix or a tendency to form A-A-A... or B-B-B...-type domains (or clusters) in the disordered matrix. Under the action of a driving force, e.g. by variation of the temperature, the disordered "matrix" structure may become thermodynamically unstable. In the former case this process may give rise to "out-of-step" domains or to completely segregated parts with – somewhat hypothetically- only one residual "domain" boundary between an A-type and a B-type crystallite. A definition of a domain structure, can be given by symmetry arguments or by the order parameter concept. Thus individual domains may be derived from a – possibly hypothetical – higher symmetric "aristophase" obeying the concept of symmetry groups. They are related to each other by a symmetry operation which is either a point group or a translational group or a "black and white"

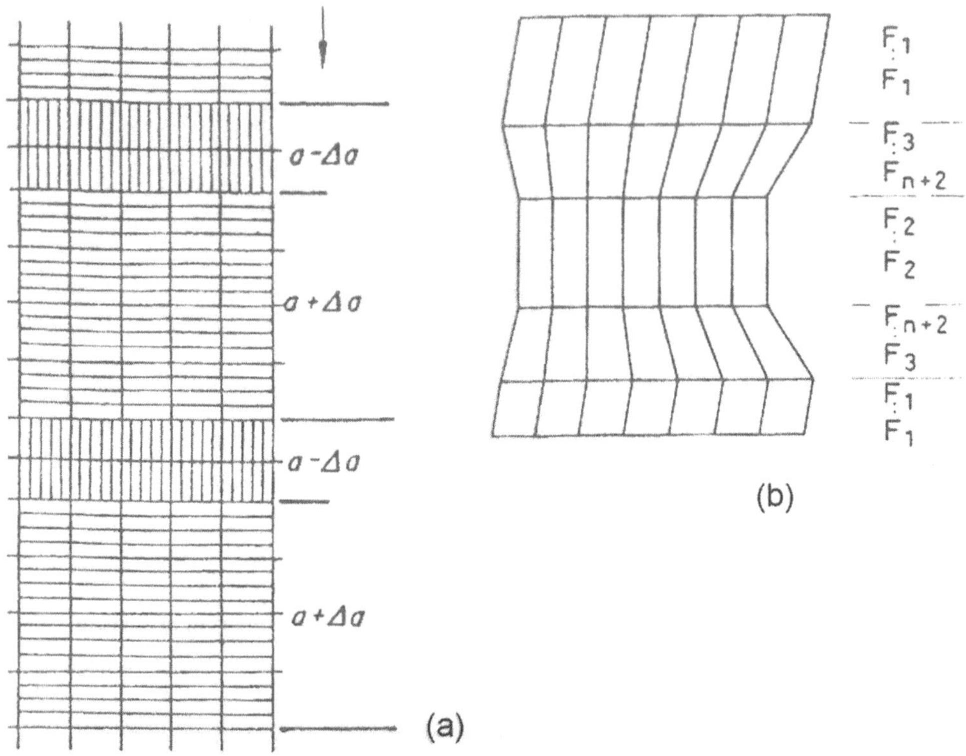

Fig. 7. Matching of domains: change of lattice constants (a), change of essential parts of the structure, giving rise to extended domain walls with gradient structures (b)

(colour) group symmetry element. The latter has to be used for magnetic domain structures and for chemical domain ordering. While preserving the same lattice, the disordered phase which is specified by "grey colour", decomposes into black and white domains, possibly embedded in the grey matrix (Fig. 3). The symmetry approach takes into account the symmetry of statistically defined structures (mixed crystals, alloys). Apparently the boundaries between the different domains are essential and must be, in principle, clearly definable. Their existence may even be used for the definition of a domain. The boundaries or domain-walls are not simple atomic planes, they are extended intermediate structural states which mellow the transition from one domain to the next one. Occasionally it might be helpful to define an extended domain wall as a new domain with a "gradient" structure which accommodates those of the neighbouring domains (Fig. 7).

On the average, this gradient structure matches the structure of the aristophase. Domain walls "carry" the symmetry change to the next domain. The strict symmetry relation between the domains may be violated as a consequence of intergrowth: misfits at the planes of coincidence may produce straining effects or single dislocations destroying an exact symmetry relation. Possibly best known are twin-domain and antiphase-domain structures. A twin is a rational intergrowth of congruent or enantiomorphic individuals of the same crystal species in two or more well-defined orientation states. Twin elements are rotational axes, mirror planes or centres of inversion. The contact interface (domain boundary) is in most cases a low-energy boundary with good structural fit. Anti-phase domains are related to a lost translation of type $1/2\boldsymbol{t_n}$ where $\boldsymbol{t_n}$ is a translational vector of the aristophase. An example is given by a binary alloy which undergoes a disorder – order transformation by cooling. Accordingly, diffracted waves suffer a phase shift of π for certain directions. More general types of "out-of-phase" domains are due to other fractional translational vectors. The example of chemical domain ordering gives us a good picture of the occurrence of real domain structures: the length scale may range from a few unit cells to mm size, i.e. macroscopic lengths. Therefore, there is no clear-cut separation between a domain structure and a new "structure" on the one side and intergrown crystallites on the other side. In other words, there is a continuous transition between the terms "disordered structure" – "domain-structure" – "polycrystal" or "microstructure". The size may be affected by many factors, for example by long-range interaction of strain-fields in ferroelastic materials [1].

Depending on the size scale of the domains, there are various methods to recognize and investigate domain structures. Apart from macroscopic methods, transmission electron microscopy (TEM) is a convenient way to study local features of a domain structure. On the other hand, the average and the distribution of domains are not easy to obtain by TEM investigations, in particular if long-range non-periodic or periodic domain structures and accompanying strain fields are present. Diffraction studies give evidence of long-range correlations in a domain structure superimposed on a crystallographic substructure by analysing more or less diffuse "reflections" [2,13,12,3,6].

3.2 Diffraction by Domains

Domain structures can be described by means of simple box functions:

$$b(\boldsymbol{r}) = 1 \quad \text{for } -A \leq r \leq +A\,(A = \alpha\boldsymbol{a}_1 + \beta\boldsymbol{a}_2 + \gamma\boldsymbol{a}_3) \quad \text{and}$$

$$b(\boldsymbol{r}) = 0 \quad \text{elsewhere, } \alpha, \beta, \gamma \quad \text{are scalar quantities}$$

$$b(\boldsymbol{r}) \leftrightarrow B(\boldsymbol{Q}) = \{\sin(\pi\alpha h)/(\pi h)\}\{..\beta..\}\{..\gamma..\}.$$

$$p(\boldsymbol{r}) = b(\boldsymbol{r}) * b(-\boldsymbol{r}) \leftrightarrow \{\sin(\pi\alpha h)/(\pi h)\}^2\{\sin(\pi\beta k)/(\beta k)\}^2\{\sin(\pi\gamma l)/(\gamma l)\}^2.$$

A general expression may then be written as:

$$\rho_D = [\rho_1 * l_1][\sum b_{1j} * d_{1j}] + [\rho_2 * l_2][\sum b_{2j} * d_{2j}] + [\rho_3 * l_3][\sum b_{3j} * d_{3j}] + \dots. \quad (19)$$

The l_i, b_{ij}, and d_{ij} describe the individual lattices in domains of type i, the shape of domain i with size j, and the distribution of domains i with size j, respectively. Clearly, the b_{ij} include a possible domain-size distribution of a domain of type i. Domain walls may be included as an own domain type i, if feasible, and may even be split up into a sequence of "domains" in case of a complicated wall structure.

Fourier transformation gives:

$$F_D = [F_1 L_1] * [\sum B_{1j} D_{1j}] + [F_2 L_2] * [\sum B_{2j} D_{2j}] + [F_3 L_3] \\ *[\sum B_{3j} D_{3j}] + \dots \tag{20}$$

and the intensity is given by:

$$I_D = \sum_i |[F_i L_i] * [\sum_j B_{ij} D_{ij}]|^2 \\ +2 \sum_{i \neq k} [F_i L_i] * [\sum_j B_{ij} D_{ij}][F_k^+ L_k^+] * [\sum_j B_{kj}^+ D_{kj}^+]. \tag{21}$$

A further mathematical treatment of this very general expression is not feasible, if no additional information is available from crystal chemistry or from experimental observations. Only some qualitative conclusions can be drawn. The first term is usually the leading term, while the second one provides, in general, only smaller additional changes. The first term corresponds to sharp or diffuse reflections which are not only determined by structure factors F_i, but also by (the Fourier transforms of) the shape, size and distribution functions. If the domain distribution is coherent with the underlying basic lattice, the D_{ij} define a set of δ-functions, (multiplied with the B_{ij}), i.e. sharp reflections (first sum in (21)), which may be accompanied by diffuse or satellite scattering. If the domain ordering process is not three dimensional, the "reflections" may be extended in streaks or planes which are (in another meaning) "diffuse" in reciprocal space. If the d_{ij} are incoherent, the cross terms are averaged out, but now the first term describes diffuse phenomena (e.g. by domain size fluctuations). Interferences between domains of different kinds may be responsible for additional extinction rules of sharp and diffuse reflections. A more detailed discussion and related references can be found in [12].

3.3 Basic Domain Structures

In the following different domain arrangements with increasing complexity are considered to learn about the diffraction features of various domain-ordered structures.

1. Two domain types with coherent lattices and random domain distribution,

   Type 1: $\rho_1(\boldsymbol{r}), l(\boldsymbol{r}) * \rho_1(\boldsymbol{r}); F_1(\boldsymbol{Q}), L(\boldsymbol{Q}) \cdot F_1(\boldsymbol{Q})$

   Type 2: $\rho_2(\boldsymbol{r}), l(\boldsymbol{r}) * \rho_2(\boldsymbol{r}); F_2(\boldsymbol{Q}), L(\boldsymbol{Q}) \cdot F_2(\boldsymbol{Q})$.

   Both domain types occur with same a-priori probability and a random mixture is assumed.

   Domain structure: $\rho_D = [l(\boldsymbol{r}) * \rho_1(\boldsymbol{r})]b(\boldsymbol{r}) + [l(\boldsymbol{r}) * \rho_2(\boldsymbol{r})](1 - b(\boldsymbol{r}))$, (22)

average structure: $\langle\rho\rangle = 1/2(\rho_1 + \rho_2); \langle F\rangle = 1/2(F_1 + F_2)$,

difference structure: $\Delta\rho = 1/2(\rho_1 - \rho_2); \Delta F = 1/2(F_1 + F_2)$.

With $b'(\boldsymbol{r}) = 2b(\boldsymbol{r}) - 1$ (b' symmetric), $b(\boldsymbol{r}) = 1/2(1 + b'(\boldsymbol{r})); (1 - b(\boldsymbol{r})) = 1/2(1 - b'(\boldsymbol{r}))$ the domain structure can be written as:

$$\begin{aligned}\rho &= [l(\boldsymbol{r}) * (\langle\rho\rangle + \Delta\rho)]1/2(1 + b'(\boldsymbol{r})) + l(\boldsymbol{r}) \\ &\quad *(\langle\rho\rangle - \Delta\rho)1/2(1 - b'(\boldsymbol{r})) \\ &= l(\boldsymbol{r}) * \langle\rho\rangle + l(\boldsymbol{r}) * \Delta\rho\, b'(\boldsymbol{r}). \end{aligned} \tag{23}$$

Fourier transformation gives:

$$F(\boldsymbol{Q}) = L(\boldsymbol{Q})\langle F(\boldsymbol{Q})\rangle + \{L(\boldsymbol{Q})\Delta F(\boldsymbol{Q})\} * B'(\boldsymbol{Q}), \tag{24}$$

$$B'(\boldsymbol{Q})) = 2B(\boldsymbol{Q}) - \delta(\boldsymbol{Q}). \tag{25}$$

There are superimposed sharp and diffuse contributions at any reciprocal lattice point, where the sharp reflections are governed by the average structure factor, the diffuse ones by the difference structure, and the "diffuseness" is due to convolution with B' (more precisely: $B'^2(\boldsymbol{Q})$). The general appearance depends on the $\Delta F/\langle F\rangle$ ratio. If there is no average structure $\langle\rho\rangle$, $\langle F\rangle = 0$, only diffuse maxima centred at the reciprocal lattice points can be observed. This latter statement could be violated at some points in reciprocal space: If the structure factor ΔF shows, occasionally, a steep gradient at a position close to a reciprocal lattice point, the convolution product $[L\Delta F] * B$ may have its maximum not exactly at a lattice point. The diffuse maximum occurs "surprisingly" at an incommensurate position which has, however, nothing to do with an incommensurate structure. (Note: This is a quite general aspect of the position of diffuse reflections which should also be taken into account if "wrong" positions or strange profiles are observed in powder patterns).

2. Distribution of one domain type in a matrix structure with coherent lattice: In a matrix structure $\rho_1(\boldsymbol{r})$, domains with a different structure $\rho_2(\boldsymbol{r})$, but the same coherent lattice $l(\boldsymbol{r})$ are embedded (see example 1). It is assumed that the domains have approximately equal sizes described again by the shape function $b(\boldsymbol{r})$. The distribution of the domains is denoted by $\boldsymbol{t_m} = m_1\boldsymbol{a_1} + m_2\boldsymbol{a_2} + m_3\boldsymbol{a_3}$ at the centres of the domains:

$$d(\boldsymbol{r}) = \sum_m \delta(\boldsymbol{r} - \boldsymbol{t_m}).$$

The total structure can be written as:

$$\begin{aligned}\rho_{tot}(\boldsymbol{r}) &= l(\boldsymbol{r}) * \rho_1(\boldsymbol{r}) - [l(\boldsymbol{r})b(\boldsymbol{r})] * \rho_1(\boldsymbol{r}) * d(\boldsymbol{r}) + [l(\boldsymbol{r})b(\boldsymbol{r})] \\ &\quad *\rho_2(\boldsymbol{r}) * d(\boldsymbol{r}), \qquad (26) \\ &= l(\boldsymbol{r}) * \rho_1(\boldsymbol{r}) - [l(\boldsymbol{r})b(\boldsymbol{r})] * [\rho_2(\boldsymbol{r}) - \rho_1(\boldsymbol{r})] * d(\boldsymbol{r}), \qquad (27)\end{aligned}$$

$\rho_1(\boldsymbol{r}) = \rho(\boldsymbol{r}); \rho_2(\boldsymbol{r}) - \rho_1(\boldsymbol{r}) = \Delta\rho(\boldsymbol{r})$, $F(\boldsymbol{Q}), \Delta F(\boldsymbol{Q})$, correspondingly:

$$F_{tot}(\boldsymbol{Q}) = L(\boldsymbol{Q})F(\boldsymbol{Q}) + [L(\boldsymbol{Q}) * B(\boldsymbol{Q})]\Delta F(\boldsymbol{Q})D(\boldsymbol{Q}), \tag{28}$$

$$\begin{aligned} I(\boldsymbol{Q}) \sim |F(\boldsymbol{Q})|^2 &= |L(\boldsymbol{Q})F(\boldsymbol{Q})|^2 \\ &+\{[L(\boldsymbol{Q})F(\boldsymbol{Q})][L(\boldsymbol{Q}) * B(\boldsymbol{Q})]\Delta F(\boldsymbol{Q})D(\boldsymbol{Q}),\} \\ &+\{cc\} + |[L(\boldsymbol{Q}) * B(\boldsymbol{Q})]\Delta F(\boldsymbol{Q})D(\boldsymbol{Q})|^2. \end{aligned} \tag{29}$$

Term 1 in eq. 29 denotes sharp reflections governed by the basic structure, terms 2 and 3 also denote contributions to the sharp reflections (multiplication with $L(\boldsymbol{Q})$, whereas term 4 describes diffuse scattering. Back-transformation of term 4 gives with $l(\boldsymbol{r}) = l(-\boldsymbol{r})$ and under the assumption of centrosymmetric domains, $b(\boldsymbol{r}) = b(-\boldsymbol{r})$, which is realistic in most cases:

$$[l(\boldsymbol{r})b(\boldsymbol{r})] * [l(\boldsymbol{r})b(\boldsymbol{r})] * \Delta\rho(\boldsymbol{r}) * \Delta\rho(-\boldsymbol{r}) * d(\boldsymbol{r}) * d(-\boldsymbol{r}). \tag{30}$$

The first convolution product in (30) gives simply $l(\boldsymbol{r})p(\boldsymbol{r})(p(\boldsymbol{r}) = b(\boldsymbol{r}) * b(\boldsymbol{r}))$. The difference structure $\Delta\rho(\boldsymbol{r})$ is a non-periodic function. We ignore $\Delta\rho$ in the following to study the influence of $d(\boldsymbol{r})$ only. With $D(\boldsymbol{r}) = d(\boldsymbol{r}) * d(-\boldsymbol{r})$ term 4 is written:

$$l(\boldsymbol{r})[p(\boldsymbol{r}) * D(\boldsymbol{r})] = l(\boldsymbol{r})D'(\boldsymbol{r}).$$

The diffuse intensity distribution is governed by the Patterson function of the distribution:

$$|L(\boldsymbol{Q}) * D'(\boldsymbol{Q})|^2$$

It depends now on the specific distribution $d(\boldsymbol{r})$ to analyse the problem further. Limiting cases may be discussed under the assumptions that the domain sizes are more or less well defined (small fluctuations) and that the perfect lattice extends far beyond the correlation length of the distribution $d(\boldsymbol{r})$. Moreover, we know that $d(\boldsymbol{r})$ cannot be completely random because the minimum distance between two domains is given by the domain size.

3. Domains with same structures and out-of-phase lattices:
Assuming a fault vector $\boldsymbol{T}$, this situation can be described by a modified expression for the domain structure discussed in example 1 ($\rho_1 = \rho_2$):

$$\{l(\boldsymbol{r}) * \rho(\boldsymbol{r})\}1/2(1 + b'(\boldsymbol{r})) + \{l(\boldsymbol{r} - \boldsymbol{T}) * \rho(\boldsymbol{r})\}1/2(1 - b'(\boldsymbol{r})) = \tag{31}$$

$$[l(\boldsymbol{r}) + l(\boldsymbol{r} - \boldsymbol{T})] * \rho/2 + \{[l(\boldsymbol{r}) - l(\boldsymbol{r} - \boldsymbol{T})] * \rho/2\}b'(\boldsymbol{r}). \tag{32}$$

Fourier transformation gives:

$$\begin{aligned} &[L(\boldsymbol{Q})(1 + \exp\{2\pi i\boldsymbol{TQ}\})]F(\boldsymbol{Q})/2 \\ &+[L(\boldsymbol{Q})(1 - \exp\{2\pi i\boldsymbol{TQ}\})]F(\boldsymbol{Q})/2 * B'(\boldsymbol{Q}). \end{aligned} \tag{33}$$

There are again sharp and diffuse peaks.

4. Strained domains:
If straining is present in a domain or a domain wall (which may also be treated as a separate domain), the lattice inside such a domain can be written as [3]:

$$ls(\boldsymbol{r}) = \sum_n \delta(\boldsymbol{r} - \boldsymbol{t_n} - \boldsymbol{s}(\boldsymbol{n}))$$

where $s(n)$ describes the atomic displacements in cell n. Fourier transformation gives

$$L_s(\boldsymbol{Q}) = \sum_n \exp\{2\pi i \boldsymbol{t_n} \boldsymbol{Q}\} \exp\{2\pi i \boldsymbol{s}(\boldsymbol{n}) \boldsymbol{Q}\}.$$

For particular cases, such as a harmonic strain modulation $\boldsymbol{s}(\boldsymbol{n})$, theories of modulated crystal structures can be applied. This aspect is however beyond the scope of this article. It should only be emphasised that, in contrast to a domain size effect, the diffuse intensity increases with increasing value of $\boldsymbol{Q}$ at cost of the accompanying reflection. Here "disorder" means fluctuations of phase and amplitude of an s-wave, or, in the case of more complicated strain modulations, the consideration of a Fourier series and corresponding superimposed sets of satellites. For long wavelengths of the strain modulations, the particular diffuse phenomena/satellites are observed close to or even within the tails of the Bragg reflections. An adequate treatment demands a detailed line profile analysis up to high $\boldsymbol{Q}$-values. Measurement of the widths of diffuse phenomena along different directions give an estimate of the anisotropic size of the domains. In practice, however, one has to take into account other diffuse intensity contributions (e.g. TDS) which may obscure the profiles.

5. Domains with non-coherent lattices:
If there are matching problems at domain boundaries or domain size fluctuations, the perfect lattice function exists only over short distances. The centres of the domains are no longer defined by one lattice function, i.e. we have a basically non-periodic d($\boldsymbol{r}$) function and a more continuous Patterson function D($\boldsymbol{r}$). Fourier transformation yields then a major contribution only to the peak $I(0)$, i.e. to the small-angle scattering regime.

4 Layered Structures, Lamellar Domains

4.1 General Remarks

One-dimensional disorder relates to structures with extended, more or less planar defects, or layered structures where "slabs" are stacked upon another with some irregularity. Most common are stacking faults. Here "disorder" relates to pairs of larger structural sub-units. To this class belong crystal structures with high anisotropy of chemical bonding, such as graphite or other layered carbon structures, layered silicate structures and transition metal dichalcogenides. The same diffraction concepts can also be used to treat lamellar type domains in metals and alloys, lamellar type segregations, extended planar domain walls, graphite-intercalation compounds, or, more generally, in domain structures of mixed crystals and inorganic compounds with impurity atoms or other small defects (vacancies, interstitials). In the latter case the formation of extended lamellar type substructures may be easily understood: if one kind of defect dominates, there is a chance of a slightly "favoured" subcell close to this defect. Even in the case of small defect concentrations, there will be a competition between the correct sequence of subcells and a minimum free energy of the defect subcell. They may have a tendency to avoid one another (in contrast to a tendency to cluster which is not of interest here). If so, wrong subcells prefer to cumulate in planes, thus minimising the surface free energy of the wrong subcell. In consequence planar boundaries are formed. Moreover, due to the tendency to avoid one another, stacking faults arrange in more or less equidistantly spaced planes, thus providing an average (super-)periodicity of lamellar domain stacking.

General diffraction phenomena are diffuse streaks or anisotropically broadened reflections parallel to the stacking direction of the layers or lamellae. Note that there is some similarity with the surface-related so-called truncation rods which have their origin in the half-infinity of the crystal, or, more precisely, the Fourier transforms of the θ-function, $\theta = 1$ and 0, inside and outside the crystal, respectively. Also the scattering of (epitaxial) artificial hetero-layer structures is "streak-like" along the normal direction. The intensity variation along these streaks allows for the analysis of the sequential ordering of the layers, their roughness etc. This paragraph is restricted to bulk-related effects in layered structures only.

4.2 Qualitative Consideration

Starting from a very simple qualitative consideration one may learn about important general results [19,6]: Let us define a basic orthorhombic cell with lattice constants $\boldsymbol{a_i} (i = 1,2,3)$. The cell is now subdivided into subcells with $\boldsymbol{a'_i} = \boldsymbol{a_i}/m_i$, where m_i are integers. For sake of simplicity a faultvector $\boldsymbol{T}$ is assumed with two components, one parallel to the $\boldsymbol{a_1}, \boldsymbol{a_2}$ plane, and one perpendicular to it, i.e. parallel to $\boldsymbol{a_3}$ (Fig. 8):

$$T = s_1 \cdot \boldsymbol{a'_1} + s_3 \cdot \boldsymbol{a'_3} = s_1/m_1 \cdot \boldsymbol{a_1} + s_3/m_3 \cdot \boldsymbol{a_3}$$

(s_i are integers with $0 \leq s_i \leq m_i - 1$).

The average domain thickness (measured along $\boldsymbol{a_3}$) is given by N_3 which is small compared to N_1 and N_2. N_i denotes the number of cells in the respective direction $\boldsymbol{a_i}$. The crystal consists of P domains where $(P \cdot N_3)$ is of the same order as N_1, N_2, i.e. P is assumed to be large. The origin of a domain p $(p = 1, 2, ..P)$ can be chosen at $(p \cdot \boldsymbol{T} + p \cdot N_3 \boldsymbol{a_3}) = p \cdot \; [s_1/m_1 \boldsymbol{a_1} + [N_3 + (s_3/m_3)\boldsymbol{a_3}]]$.

With $\boldsymbol{t_n} = n_1 \boldsymbol{a_1} + n_2 \boldsymbol{a_2} + n_3 \boldsymbol{a_3}$ the scattered amplitude $A(\boldsymbol{Q})$ can be written as

$$A(\boldsymbol{Q}) = F(\boldsymbol{Q}) \cdot \sum\sum\sum\sum \exp\{2\pi i(\boldsymbol{T} + \boldsymbol{t_n})\boldsymbol{Q}\}$$

where the summations over n_1, n_2, n_3, and p extend from 0 to $(N_1 - 1), (N_2 - 1), (N_3 - 1), P - 1$, respectively. We end up with 3 summations which may be

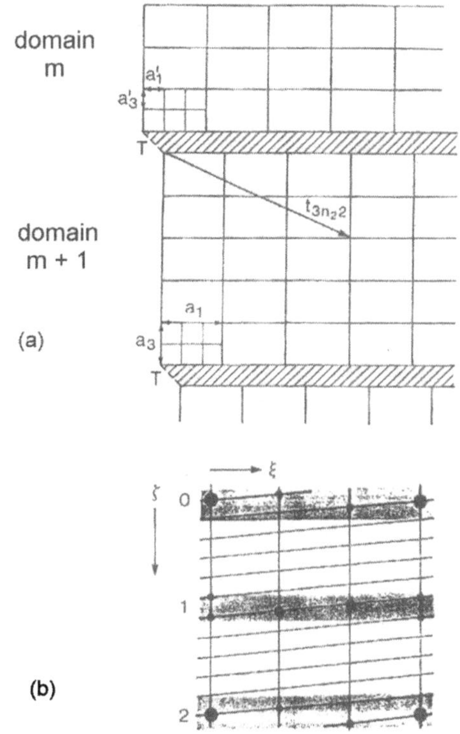

Fig. 8. Schematic picture of domain disorder with $\boldsymbol{T} = \frac{1}{3}\boldsymbol{a_1} + \frac{1}{2}\boldsymbol{a_3}$ (a), and corresponding diffraction pattern (b)

approximated by δ -functions for large N_1, N_2, and P, and a fourth term which is a Laue function L_4 due to the finite lattice:

$$A(\boldsymbol{Q}) = G(\xi) \cdot G(\eta) G(\xi, \zeta) L_4(\zeta),$$

$$G(\xi) = \sum \delta(\xi - h); G(\eta) = \sum \delta(\eta - k),$$

$$G(\xi, \zeta) = \sum \delta([s_1/m_1 \cdot \xi + (N_3 + s_3/m_3) \cdot \zeta] - m),$$

$L_4(\zeta) = \{\sin(\pi N_3 \zeta)/\sin(\pi\, \zeta)\}$ which has maxima at positions $\zeta = \ell$

($h, k, \ell, m =$ integers; $\xi, \eta, \zeta =$ fractional coordinates between 0 and 1).

Apart from the influence of the structure factor we have strong and weak reflections, depending on the variation of L_4. The weak ones are asymmetrically positioned close to the positions of the reciprocal lattice points belonging to the structure of a domain (Fig. 8b). The special case $s_1 = 0$, $s_3 = 1$, $m_3 = 2$ describes an anti-phase fault vector (perpendicular to the lamellae) giving rise to (broadened) satellite vectors close to the (extinct) l = odd positions. Due to domain-size fluctuations and the restricted correlation lengths (along $\boldsymbol{a}_3$), the weak reflections become more and more diffuse and may coalesce in one diffuse maximum or a short diffuse streak along $\boldsymbol{a}_3$.

Along the same simple treatment diffraction phenomena of disordered lamellar domains with coplanar fault vectors can be derived qualitatively. Assuming a disorder with respect to only two, e.g. "split" positions, for example $\pm s \cdot \boldsymbol{a_1}$ or $\alpha \cdot s\, \boldsymbol{a_1} (\alpha = \pm 1)$, it is possible to describe twin domains or polytypic domains , or any other type of disordered domains. The origin of a domain which is p periods away from a reference layer is given by

$$\boldsymbol{T} = s \cdot \sum_j \alpha_j \cdot \boldsymbol{a_1} + n_1\, \boldsymbol{a}_1 + n_2\, \boldsymbol{a}_2 + p\, \boldsymbol{a}_3.$$

If there is no further conditional probability, the values of α fluctuate randomly from one layer to the next. The $\sum \alpha_j$ is even or odd, if p is even or odd, respectively. This can be expressed by

$$\sum \alpha_j = p + 2z_p$$

with -p$\leq z_p \leq 0$.

The scattered amplitude of a crystal consisting of P domains can then be derived

$$A(\boldsymbol{Q}) = F(\boldsymbol{Q}) \sum \sum \exp\{2\pi i(n_1\xi + n_2\eta)\} \exp\{2\pi i[2 \cdot s \cdot z_p\, \ell + p(s\xi + \zeta)]\}. \quad (34)$$

The first two summations may again be well approximated by δ-functions $\delta(\xi - h)$, $\delta(\eta - k)$(h,k = integers), the last term depends on the value of s.

As an example we take s = 1/4, which is observed in several compounds: for $h = 4n$ $(n = 0, \pm 1, \pm 2,) : 2 \cdot s \cdot z_p \cdot h =$ integer, $s \cdot h=$ integer; ....

$$\sum \exp\{2\pi i \cdot p\zeta\} \quad \text{shows peaks for} \quad \zeta = integer,$$

for $h = 4n \pm 2 : 2 \cdot s \cdot z_p \cdot h =$ integer $s \cdot h =$ integer $/2 (= m/2)$

$$\sum \exp\{2\pi i \cdot p \cdot (n/2 + \zeta)\} \quad \text{shows peaks for} \quad \zeta = m/2.$$

For h = odd: no general conclusion is possible, as z_p is unknown. However, the sum will vanish only at particular ζ-values and diffuse streaking parallel to $\boldsymbol{a}_3^*$ through these reflection positions will most likely occur.

Another example is the formation of lamellar domains by stacking hexagonal close-packed planes along $\boldsymbol{a}_3$ $(\boldsymbol{a}_1, \boldsymbol{a}_2, \boldsymbol{a}_3$ denote the basis in a hexagonal Bravais lattice) with erratic or more cooperative faulting by vectors $\pm(1/3 \cdot \boldsymbol{a_1} + 2/3 \cdot \boldsymbol{a_2})$. Sharp reflections occur at positions (h+k) = 3n (n=integer, ζ = integer), those with (h+k) = 3n ± 1 are more or less diffuse, even streak-like. Note that such layers occur in all kinds of close-packed structures, including those with a cubic fcc Bravais lattice.

4.3 Quantitative Treatments

Close-packed structures belong to the earliest examples that were studied by quantitative analysis of diffuse scattering. A summary of these early papers can be found in [11]. Typical diffuse phenomena are streaks parallel to the normal of the layers. In a very simple model of stacking of only one kind of layer, with only nearest neighbour interactions, stacking faults are included viz. displacements vectors $\boldsymbol{t}$ parallel to the basal plane [11]. With a certain probability α, $\boldsymbol{t}$ assumes values which are not those of a lattice vector, thus giving rise to "wrong" positions. Therefore, depending on the direction and the length of $\boldsymbol{t}$, some of the reflections remain sharp, while others become streaky, i.e. diffuse along $\boldsymbol{c}^*$. The intensity is given by:

$$I(\ell) = N \cdot |F|^2 \cdot \{1 - (1-2\alpha)^2\}/\{1 - (2-2\alpha)\cos 2\pi\ell + (1-2\alpha)^2\}. \quad (35)$$

N denotes the number of layers, F the structure factor of one unit cell of an unshifted layer. Hendricks and Teller [29] derived an expression for stacking faults in close-packed structures. This formula was successfully applied for the interpretation of diffuse scattering by stacking faults in Rb-intercalated pyrolytic graphite $C_{24}Rb$ [30]. Defining an interplanar correlation parameter x, the intensity distribution along the streak direction I(ℓ), or I(Φ) , where $\Phi = 2\pi\ell/3$, can be written as:

$$\begin{aligned} I(\Phi) &= \frac{(1+x)^2}{(4-3x^2)^{1/2}} \\ &\cdot \left\{ \frac{(4-3x^2)^{1/2}\,(2-\cos\Phi/2) + 3\,i\,x\,\sin\Phi/2 - 2 + 4\cos\Phi/2}{2(1+x)^2 - (2-x^2-x(4-3x^2)^{1/2})\cdot\exp\{-i\Phi\}} \right\} \\ &+ \left\{ \frac{(4-3x^2)^{1/2}(2-\cos\Phi/2) - 3ix\sin\Phi/2 + 2 - 4\cos\Phi/2}{2(1+x)^2 - (2-x^2+x(4-3x^2)^{1/2})\cdot\exp\{-i\Phi\}} \right\} \\ &+ \{cc\} - (2-\cos\Phi/2). \end{aligned} \quad (36)$$

As a result, uncorrelated stacking faults of [C-Rb-C]-sandwiches were found to occur with a probability of 20%.

In more sophisticated approaches next-but-one and more interactions between the layers have to be considered. This was done in the so-called difference equation method by introducing a "Reichweite"-parameter s (≥ 1) [14] or in the matrix method [17]. Both methods are largely equivalent [15]. More kinds of layers may also be introduced. The general form of the final formulae is similar to (35) where terms (1- α) have to be replaced by the parameters λ_j, which are the eigenvalues of the relevant characteristic equation. The inherent difficulty lies in the extraction of the a-priori and a-posteriori (= conditional) probabilities from the experimentally determined parameters λ_j. There are several approaches and particular cases related to "deformation faults", "twin faults", "polytypic" variants,... which are not discussed here in detail. As a general outcome there are characteristic reflection profiles and intensity ratios of the diffuse contributions relative to Bragg intensities. A very general expression depends on inherent asymmetries of the layer statistics [14]. The formula includes a continuous transition from sharp (Bragg) reflections to a rod-like diffuse intensity distribution with asymmetric profiles. As a simple example the case of two types of lamellae with only next nearest neighbour interactions is briefly considered [16]. The lamellae have structures ρ_1, ρ_2 with corresponding structure factors F_1, F_2, and lattice constants $\boldsymbol{a_3} \pm \boldsymbol{\Delta}$, but a coherent plane of intergrowth:

$$I(\zeta) = \sum_{j=1,2} \left[\frac{A_j \cdot (1-|\lambda_j|^2)}{1-2|\lambda| \cdot \cos(2\pi\zeta + \Phi_j) + |\lambda_j|^2}\right] - \left[\frac{2B_j|\lambda_j| \cdot \sin(2\pi\zeta + \Phi_j)}{1-2|\lambda_j|\cos(2\pi\zeta + \Phi_j) + |\lambda_j|^2}\right]. \quad (37)$$

λ_j and Φ_j are determined by $\lambda_j = |\lambda_j| \exp\{2\pi i \Phi_j\}$ where λ_j are the solutions of a quadratic equation:

$$\lambda^2 - \lambda(\alpha \cdot \exp\{2\pi i \boldsymbol{Q\Delta}\} + \beta \exp\{-2\pi i \boldsymbol{Q\Delta}\}) - 1 + \alpha + \beta = 0.$$

The A_j and B_j represent real and imaginary parts of constants which can be calculated from the structure factors F_i and the inherent probabilities α and β (and also some other phase factors) in a cumbersome way. Each λ_j describes a reflection: $m = 1$ and 0 describe the limiting situation of a sharp reflection and a diffuse streak. Depending on the values of α, β, and Δ, different contributions of sharp and superimposed diffuse reflections may be evaluated. The first term in (37) governs the integrated diffuse intensity, the second one relates to possible asymmetric profiles. The situation of an incoherent plane of intergrowth is much more complicated.

For the case of stacking faults of hexagonal closed packed layers the characteristic distribution of sharp and diffuse reflections which was mentioned above, is also derived. In an extended form of (37), some more terms have to be included which become significant only if the number of affected layers N is small. This general formula was applied to disordered intermediate structures in course of the martensitic fcc $\leftrightarrow$ hcp phase transformation in pure Co [7], and was also used to interpret stacking faults of layers consisting of bucky-balls in supramolecular C_{70} fullerites [31]. A further extension includes impurity atoms and their influence on the stacking disorder. By use of Monte-Carlo-simulation techniques it could be shown that a basic consequence of the presence of Cu impurities in Co is an additional scattering along rows in reciprocal space which are free from diffuse scattering according to the classical theories [32].

There are many more similar theories of that kind describing one-dimensional disorder of layered type structures which will not be given here. A generally different approach is the direct calculation of the Patterson (or auto-correlation) function of a disordered layer structure [2]. Rather lengthy expressions are derived for an arbitrary number of different types of layers and corresponding characteristic probabilities for particular types and particular distances of the layers as well as general displacement vectors $\boldsymbol{t}$. They describe the distribution of sharp and streaky diffuse reflections. Expressions are given in [2]. Summarising here only the simplest case of random faulting of identical layers related by displacement vectors $\boldsymbol{t}$ parallel to the layers and occurring with probabilities α, the disordered structure is described by the convolution product $\rho_o * d(\boldsymbol{r})$ where ρ_o denotes the structure of one layer and $d(\boldsymbol{r})$ some distribution. The Patterson function is $P(\boldsymbol{r}) = \rho_o(\boldsymbol{r}) * \rho_o^+(-\boldsymbol{r}) * D(\boldsymbol{r})$. D describes the probability that if one layer is centered at the origin, another one will be centered at $\boldsymbol{r}$. Using Poisson distribution functions with a parameter α, the following intensity expression is derived:

$$I(\boldsymbol{Q}) = |F(\boldsymbol{Q})|^2 \cdot \frac{1 - \exp\{2\alpha c - 1\}}{1 + \exp\{2\alpha c - 1\} - 2\exp\{\alpha c - 1\} \cdot \cos(2\pi \boldsymbol{Q} \cdot \boldsymbol{t_n} + \alpha \cdot s)} \quad (38)$$

where $c = \cos(2\pi \boldsymbol{Q} \cdot \boldsymbol{t})$ and $s = \sin(2\pi \boldsymbol{Q} \cdot \boldsymbol{t})$.

Instead of using a large number of probabilities to describe increasing domain sizes, domain size distributions are introduced which can be described by different functions [21]. A domain of type k has a structure ρ_k (with corresponding Fourier transform F_k) and a lattice is defined by $\boldsymbol{a_{1k}}, \boldsymbol{a_2}, \boldsymbol{a_3}$. Different domains of same type k have different sizes $\boldsymbol{a_{1k}} \cdot A_k$. These sizes have distributions $w_k\ (A)$. Different size distributions are assumed to be statistically independent. The coherent planes of intergrowth are assumed to be parallel to $\boldsymbol{a_2}$ and $\boldsymbol{a_3}$, the streak-like diffuse scattering occurs therefore perpendicular to these planes. The crystal is made up by a cyclic sequence of groups of k different types of domains. Each group is counted by ℓ , the total number of domains is L_D. The origin of the lattice G_{kl} of domain D_{kl} (type k, group ℓ)is displaced against a reference origin by a fault vector $\boldsymbol{T}_k$. Within the frame of this formalism the domain structure can be written as:

$$\rho \cdot d_D = \sum_k \sum_l G_{kl}(\boldsymbol{r} - \boldsymbol{r_{kl}})$$

where $\boldsymbol{r_{kl}}$ denote the origins of each domain kl, the summations extend from 0 to A_{kl} which is the width of domain kl. An intensity expression is derived by using the Patterson function:

$$I(\boldsymbol{Q}) = P_s(\boldsymbol{Q}) + 2\{\mathcal{R}\}[1/(1 - R(\boldsymbol{Q})) \cdot P_d(\boldsymbol{Q})(\{\mathcal{R}\} = \text{ real part of } [...]). \quad (39)$$

P_s denotes the Fourier transform of the average Patterson function of uncorrelated domains which governs the sharp contribution of the diffraction. The more complex expressions P_d include the dependences of the (average) domain sizes B_k, lattices G_k, structure factors F_k and fault vectors T_k. R($\boldsymbol{Q}$) is an abbreviation for the product of all $B_1 \cdot B_2 \cdot B_3$, .. times a phase factor $\exp\{2\pi i\ \boldsymbol{Q}(\boldsymbol{T_1} + \boldsymbol{T_2} + ..\boldsymbol{T_k})\}$. This theory was successfully applied to the case of lamellar exsolutions in lunar pyroxenes where the domains are made up by Ca-rich (augite) and Ca-poor (pigeonite) lamellae [21].

Example: Domain ordering during the phase transformation in enstatite $MgSiO_3$:

Pyroxenes are chain-silicates and form an important group of rockforming minerals. Enstatite, $MgSiO_3$, occurs in three basic ambient-pressure polymorphs

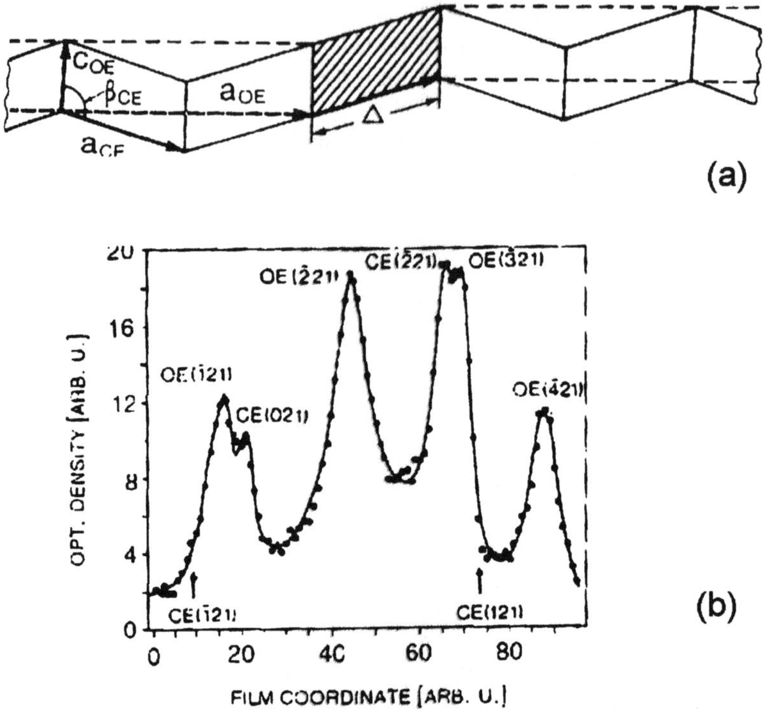

Fig. 9. Out-of-step domains in orthoenstatite, the shaded area marks a domain wall (a); diffuse intensity along the (h21)-direction in reciprocal space: profile fitted by intergrown OE and twinned CE domains. Figure taken from [4]

with reconstructive phase transformations between them. Orthorhombic protoenstatite (PE) is stable above 1300 K, while both orthorhombic orthoenstatite (OE) and monoclinic clino-enstatite (CE) exist at lower temperatures. On cooling from high temperatures (PE phase) a disordered domain structure of intergrown CE and OE domains develops [4]. The structures may be visualised as being composed of two structural layers parallel to the $\boldsymbol{a_2} - \boldsymbol{a_3}$ plane. The layers consist of SiO_4-chains and bands of MgO_6 octahedra which differ mainly in the orientation of the so-called M1 octahedra. They will be denoted by "+" and "-" symbols. The different polymorphs differ by different stackings of the layers parallel to the $\boldsymbol{a_2} - \boldsymbol{a_3}$ plane: PE: +-+-+-, CE ++++ or - - - - (twin domains) and OE ++- -++- -. OE may be understood formally as microtwinned CE on a unit cell scale. Due to slight differences of equally oriented layers in CE and OE the unit cell of CE contains two such layers: $2 \cdot a_{CE} \cdot \sin\beta \approx a_{OE}$($\beta$ = monoclinic angle). In the course of the PE–OE/CE transformation OE domains separated by CE lamellae, i.e. out-of-step domains, are formed with fault vectors $\boldsymbol{T} = \Delta_1 \boldsymbol{a_{1,OE}} + \Delta_3 \boldsymbol{a_{3,OE}}$ (Fig. 9a). This gives rise to diffuse streaking and anisotropic broadening of reflections [3,4] (Fig. 9b).

By means of the theory outlined above, (33), the sharp and diffuse contributions of the reflections can be written as

$$I_{s/d} \sim (1 \pm \cos 2\pi(\Delta_1 \cdot \xi + \Delta_3 \zeta))$$

(subscripts s and d relate to + and - sign, respectively).

It was found that reflections (h21) with h = even have a "sharp" contribution of only 7%, whereas those with h = odd have a "sharp" contribution of 93%. A comparison between measured and calculated profiles by using $\Delta_1 = 1/2$ and $\Delta_3 = a_1/a_3 \cdot \tan(\beta - \pi/2)$ was performed. The analysis of the diffuse scattering allowed for a determination of an irrational fault vector.

5 Quasicrystals

5.1 Aperiodic Order

Quasicrystals belong to the class of aperiodic crystals. Perfect aperiodicity implies strict long-range orientational order which extends, in principle, to infinity comparable to conventional crystals. Quasicrystalline structures are self-similar, corresponding diffraction patterns show scaling (inflation, deflation) symmetry related by (a function of) τ, i.e. the golden mean:

$$\tau = 2 \cdot \cos(\pi/5) = (1 + \sqrt{5}/2) = 1.618\dots .$$

In diffraction patterns a most exciting feature is the presence of non-crystallographic rotational symmetries, e.g. five- or tenfold rotations, although this aspect seems not to be a decisive criterion of quasicrystallinity [33]. Prominent examples are the "icosahedral phases" whose diffraction patterns exhibit the combination of symmetry elements of 5-fold rotations, 2-fold rotations and of mirror planes which are typical for an icosahedron, and "decagonal" phases where aperiodic and periodic order coexist: there is a unique direction along which the structural units are periodically arranged, whereas aperiodic order occurs in the remaining two dimensions perpendicular to the periodic axis.

The density of a dD aperiodic crystal can be expressed by Fourier series:

$$\rho(\boldsymbol{r}) = \sum_{Q} F(\boldsymbol{Q}) \cdot \exp\{i\boldsymbol{Q}\boldsymbol{r}\} \tag{40}$$

with

$$\boldsymbol{r} = \sum_{i=1}^{d} x_i \cdot \boldsymbol{a_i}. \tag{41}$$

$\boldsymbol{Q}$ is a wave vector that may be expressed by

$$\boldsymbol{Q} = 2\pi \sum_{j=1}^{n} h_j \cdot \boldsymbol{a_j}^*. \tag{42}$$

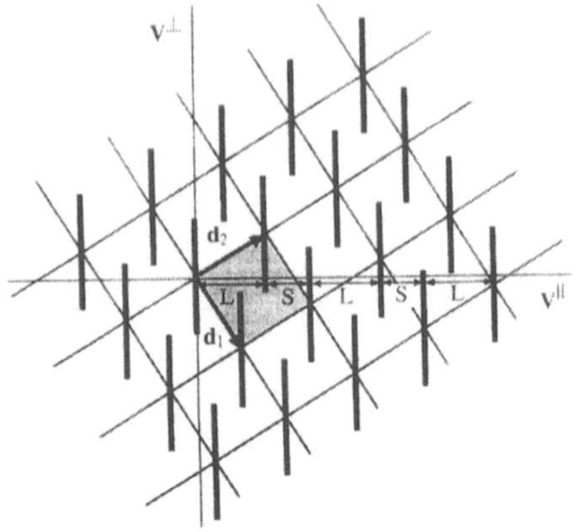

Fig. 10. Schematic picture of a 1D quasicrystal embedded in a 2D hyperspace; hyperatoms are extended along the perpendicular space; their cut with the parallel space define the real positions of atoms [27].

The indices h_i are elements of a Z-module of rank $n(> d)$, and $\boldsymbol{a_i}$ (i=1...d) and $\boldsymbol{a_j^*}$ (j = 1...n) are the basis vectors in the corresponding space. Both sets, however, are not orthogonal ("reciprocal") to each other. The diffraction pattern of an ideal aperiodic quasicrystal consists of a discrete, but "dense" Fourier spectrum with components F($\boldsymbol{Q}$).

Translational symmetry, i.e. periodicity in direct space, as well as orthogonality between the basis vectors in direct and reciprocal space, can be restored in a higher dimensional description where a minimal dimension n in direct space is introduced. The higher dimensional direct space with vectors $\boldsymbol{r}$ may be chosen in such a way that it decomposes into two orthogonal subspaces, the so-called parallel (or external or physical) subspace with dimension d, and the perpendicular (or internal or complementary) subspace with dimension (n-d). Correspondingly the vectors $\boldsymbol{r}$ may be written as:

$$\boldsymbol{r} = \sum^{d} x_i \cdot \boldsymbol{a}_i = \sum^{n} x_i \cdot \boldsymbol{d}_i = (\boldsymbol{r}_{\|}, \boldsymbol{r}_{\perp}) \tag{43}$$

and the wave vectors as

$$\boldsymbol{Q} = 2\pi \sum^{n} h_i \cdot \boldsymbol{d}_i *, h_i \in Z. \tag{44}$$

The parallel (physical) dD subspace is embedded in the nD hyperspace in such a way that the "cut" with the hyperlattice is defined by an irrational number. That means the cut goes through just one arbitrary point taken as origin (Fig. 10). The structure of the quasicrystal, i.e. the positions of the atoms in dD space, can be described by means of "hyperatoms" or "hyperatomic surfaces" which are extended in the (n-d)D perpendicular space, and remain "sharp" in physical subspace. The "cut" intersects the hyperatoms and the intersections

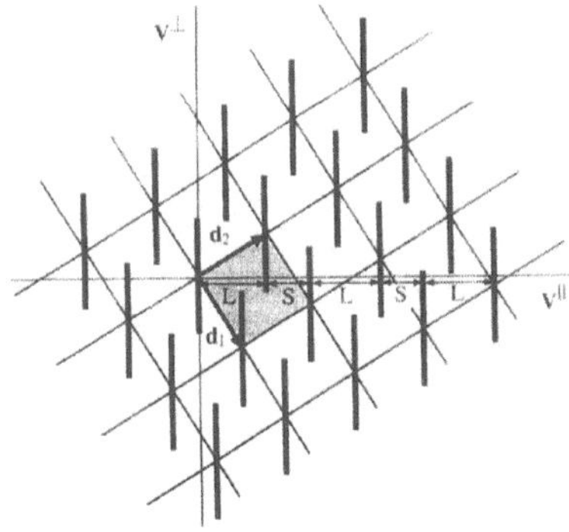

Fig. 11. Reciprocal lattice of 1D quasicrystal is embedded in 2D space. The projection of the lattice points in V^* onto the 1D subspace $V_{\parallel}^*$ gives the diffraction pattern [27]

define the positions of the real atoms in dD physical subspace. Note that the atomic surface may have parts with different chemical character. Summarising, the aperiodic dD structure is the result of an irrational cut of the physical space through an nD periodic hyperstructure at $\boldsymbol{r}_{\perp}$ =0:

$$\rho(\boldsymbol{r}_{\parallel,0}) = \sum_{Q} F(Q) \cdot \exp\{i\boldsymbol{Q}\boldsymbol{r}\}. \tag{45}$$

As a consequence, the diffraction pattern of a quasicrystal can be described as a projection of the nD periodic reciprocal hyperlattice upon the ("ordinary") dD reciprocal space of the quasicrystal (Fig. 11). The duality condition $d_i \cdot d_j^* = \delta_{ij}$ (i, j = 1...n) is restored. The vectors $\boldsymbol{d}_i$ and $\boldsymbol{d}_j^*$ are now the basis vectors of the nD hypercrystal in the dual spaces and we can safely speak about a reciprocal lattice of the quasicrystal. Apparently there are additional (n-d) "phason" degrees of freedom corresponding to the dimension of the complementary space.

5.2 Disordered Aperiodic Structures

In quasicrystals there is the chance of "phason-like" displacement disorder along directions in the complementary space, in addition to the common "phonon-like" disorder or straining in the physical subspace. In contrast to phonon-like displacements which are responsible for continuous local distortions of the atomic structure, "phason-like" displacements may describe discontinuous atomic jumps or "flips". This disorder type is unknown in conventional crystals. The behaviour may easily be visualised by displacing an atomic surface along a coordinate in internal space. It is possible that the cut intersects the atomic surface at a chemically different position or does not intersect the surface at all. As a consequence, we observe either, in former case, a chemical disorder or, in the latter case, an atom "disappearing" from its original position. An atom "reappears" at a different position, due to a cut of another atomic surface (belonging to a different hyperlattice point). There are different kinds of phason-like disorder including

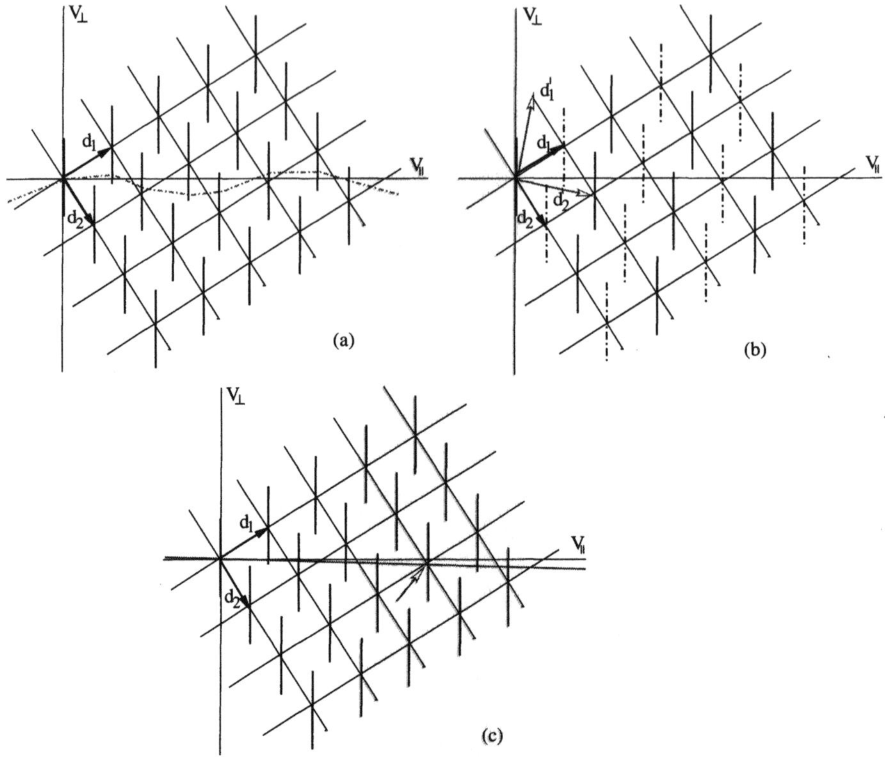

Fig. 12. Examples of phasonic disorder and super-order in the higherdimensional description: (a) random phason fluctuations; (b) chemical superordering of the hyperlattice; (c) linear phason strain: the arrow marks the lattice point cut by $V_{\|}$

random phason displacements, which play a major role in the so-called random phason tilings, or phason type modulations and phason straining, and also shape or size fluctuations of the hyperatomic surfaces. Various phasonic types of disorder are schematically shown in Fig. 12. Apart from these displacive types, quasicrystals may have a certain amount of substitutional disorder due to compositional fluctuations in the binary or ternary intermetallic phases. Deviations from a "perfect" solid solution towards chemical related structural inhomogeneities occur and give rise to complex microstructures. Another source of disordering in quasicrystals is also due to the "competition" of aperiodic and periodic order in quasicrystals as function of composition and/or temperature. Metastable quasicrystals are frequently affected by locally coexisting aperiodic and periodic domains as a consequence of rapid solidification during the growth process of the sample. Upon heating to or cooling from melting temperature, transient ordering states occur in course of phase transformations between the quasicrystalline and the crystalline state or during annealing or homogenising procedures. These order-disorder phenomena strongly depend on the exact chemical composition. The term "transient ordering states" includes not only structural modifications

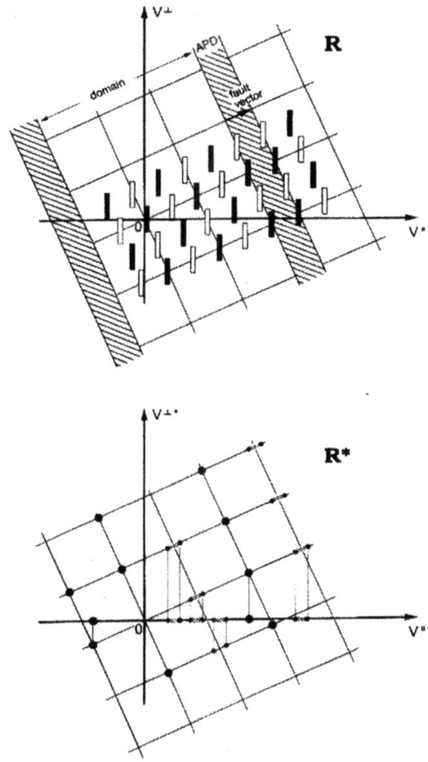

Fig. 13. Schematic of lamellar domain disorder in hyperspace and corresponding diffraction

or strained regions, but also the intergrowth of nanodomains with aperiodic or so-called 1D quasicrystalline phases or twin-related crystalline "approximant" phases. Also superordering phenomena, possibly short-ranged, may be discussed in the internal space, thus introducing super-aperiodic order or domains in superspace [8] (Fig. 13). Furthermore, complex diffuse scattering phenomena in decagonal phases are related to their "hybrid" translational ordering.

5.3 Diffraction of Disordered Quasicrystals

In the Fourier representation of a disordered quasicrystal, the expression for the scattering density includes a phase factor:

$$\exp\{i[\boldsymbol{Qr} + \Phi(\boldsymbol{Qr})]\}$$

where Φ covers phonon- ($\boldsymbol{u}$-) and phason-like ($\boldsymbol{w}$-)displacements

$$\Phi(\boldsymbol{Qr}) = \boldsymbol{Q}_{\|} \cdot \boldsymbol{u}(\boldsymbol{r}) + \boldsymbol{Q}_{\perp} \cdot \boldsymbol{w}(\boldsymbol{r}).$$

$\boldsymbol{Q}_{\|}$ and $\boldsymbol{Q}_{\perp}$ are the components of the scattering vector in the two subspaces introduced above. The $\boldsymbol{u}$ (phonon-type) or $\boldsymbol{w}$ (phason-type) fields may be anisotropic and non-linear. As in the case of conventional disordered crystals, we may try to divide the scattering density into in an average part and a fluctuational part. The average part governs the Bragg scattering, the fluctuational part concerns diffuse scattering contributions (see Sect. 2). The average structure factor writes

$$\langle F(\boldsymbol{Q})\rangle \sim \sum_k \exp\{i\boldsymbol{Q}\cdot\boldsymbol{r}_k\}\, f_k(\boldsymbol{Q}_\parallel)\, g_k(\boldsymbol{Q}_\perp) n_k \\ \exp\{-\boldsymbol{Q}_\parallel\, \boldsymbol{B}_\parallel\, \boldsymbol{Q}_\parallel^T\}\exp\{-\boldsymbol{Q}_\perp\cdot\boldsymbol{B}_\perp\cdot\boldsymbol{Q}_\perp^T\} \tag{46}$$

where f_k denotes the Fourier transform of the size of the true atomic scattering density k and is simply the well known atomic form factor; g_k denotes the Fourier transform of the shape of the average atomic hypersurface k; $n_k(\leq 1)$ is an occupation factor; the last two exponentials are Debye-Waller-factors due to the "extension", i.e. non-point-like scattering distribution in physical and perpendicular subspaces. $\boldsymbol{B}_{\parallel,\perp}$ are the mean-square displacement matrices in the respective subspaces. Both effects may be separated by analysing the diffracted intensities in dependence on $\boldsymbol{Q}_\parallel$ and $\boldsymbol{Q}_\perp$.

As in the case of conventional crystals, there is no unique theory of diffuse scattering of quasicrystals. A basic concern is the dispute whether or not quasicrystals are theoretically best described by "random tiling models" which inherently contain random phason flips. As shown in [34] peak broadening and diffuse intensity can be simulated if introducing randomness in an ideal 3D Penrose tiling model. Most of the investigations are carried out on the icosahedral (i) and decagonal (d) phases. Basic structural elements of i- phases are different types of polyhedra, in particular icosahedral clusters, which are arranged aperiodically. A phason elastic theory was developed for icosahedral quasicrystals with quenched random phason strains [35,36]. The disorder theory is based on a continuum elastic free energy and includes also phonon-phason coupling terms, and also Huang type (anisotropic) diffuse scattering close to the Bragg peaks, which is a consequence of strain fields due to quenched random defects. Diffuse intensity contours in the vicinity $\boldsymbol{q}$ of a Bragg reflection at $\boldsymbol{Q}$ are given by:

$$|\boldsymbol{q}_\parallel|^2 \sim \sum_{\alpha=1}^{6} \frac{1}{\omega_\alpha^2(\boldsymbol{q})}(\boldsymbol{Q}_\parallel e_{\parallel\alpha}(\boldsymbol{q}) + \boldsymbol{Q}_\perp e_{\perp\alpha}(\boldsymbol{q}))^2. \tag{47}$$

where $q_\parallel = (\boldsymbol{Q}-\boldsymbol{H})$, and $\boldsymbol{e}(\boldsymbol{q})_\parallel$ and $\boldsymbol{e}(\boldsymbol{q})_\perp$ as well as $\omega(\boldsymbol{q})^2$ denote the 3+3 unit eigenvectors and the 6 eigenvalues of the hydrodynamic matrix, respectively. There is an increase of PDS and TDS with $\boldsymbol{Q}_\perp^2$, and $\boldsymbol{Q}_\parallel^2$, respectively, and a decay of the PDS with $1/\boldsymbol{q}^2$, similar to the behaviour of conventional TDS in crystals. In other words, phason related diffuse scattering is always strong at positions with high components $\boldsymbol{Q}_\perp$. There is again the analogy with regular crystals where the TDS increases with the modulus of the scattering vector $\boldsymbol{Q}$. More detailed studies were undertaken by [37] and [38], where the effect of diffuse scattering due to phason softening accompanying phase transitions in i-quasicrystals was investigated. Neutron diffuse scattering of i-AlPdMn [40] was interpreted in the framework of this hydrodynamic theory using two phason elastic constants.

Relatively strong arcs of localised diffuse scattering have been observed in Ti-based i-phases [41]. This scattering was analysed in terms of an icosahedral glass model with some short range ordering. This model gives a good description of the general features, i. e. sharp peaks, localised diffuse arcs, and a broad diffuse background. This topological disorder model based on a random packing of clusters was also used by [42] for i-AlMnSi. Ring-/arc-like diffuse scattering was reported to occur in i-AlCuLi [43]: a simple expression of diffuse scattering which could be modelled by short range order of icosahedral clusters:

$$I(\boldsymbol{Q}) \sim f_{Cl}^2 \langle |1 + P_j \exp\{i\boldsymbol{Q}\boldsymbol{R_j}\}|^2 \rangle. \tag{48}$$

f_{Cl} denotes a cluster form factor, P_j gives the probability of joining another i-cluster via any of the 20 cluster faces, and $\boldsymbol{R}_j$ denotes the 20 intercluster vectors. Other disorder models in i-phases are due to vacancies, strains, dislocations or the Hume-Rothery mechanism [44].

The disorder phenomena in decagonal phases are even more complicated than those in i-phases. Significant structural elements of d-phases are columns or clusters of columns parallel to the unique "periodic" axis. For sake of brevity this axis is called c-axis in the following. A basic period along c covers two atomic layers, different periodicities which are multiples of this basic c-repeat unit are observed in different d-phases. Either a tenfold or a fivefold rotation axis along c is the most striking symmetry element. Disordering and corresponding diffuse scattering occurs with respect to the aperiodic order in two dimensions, which can be discussed in the higher dimensional description in a 4D superspace, and with respect to the remaining c-direction. Therefore, research of structural disorder in d-phases refers to a (4+1)D superspace.

Compared with the theory of phason elasticity of the i-phases, a similar theory was developed for decagonal quasicrystals [45,39]. Rather lengthy expressions for diffuse scattering are derived, which are due to long-wavelength phasonic modulations (PDS), phonons (TDS), or phason-phonon coupling terms, as well as due to elastic strains caused by interstitials, vacancies or dislocations, i.e. Huang-type diffuse scattering. For the case of d-AlNiFe a quantitative comparison with experimental results was carried out and a mixing of different contributions seems to give the best matching of experimental data and theory [46].

Cooperative aperiodic structural disorder phenomena in d-phases were studied in d-AlNiCo, d-AlCuCo and other isotypic phases. (The term isotypic refers to similar average structures of all these d-phases). As known from direct electronmicroscopical observations, there are intimate intergrowths of approximant phases or twinned aggregates of approximant phases, coexisting microcrystalline (periodic) and true quasicrystalline regions, modulated or superlattice ordered quasicrystalline phases and all kinds of transitions between the different structural states. An example of the diffuse scattering is shown in Fig. 14 [47]. The structural re- and dis-ordering occurs on a length scale of nanometers. A quite general feature of the domain formation is a lamellar type domain ordering. This can be concluded from the observation of streak-like diffuse scattering within the diffraction patterns of the quasiperiodic plane.

Fig. 14. $Al_{72.5}Co_{16.5}Ni_{11}$, [hk0] layer left, and diffuse layer line system ([h0l]) right (image plate recording)

A tentative explanation by domain formation in 4D hyperspace is given in [8]. Very common are phason- strained lamellar domains. The kind and amount of diffuse scattering depend very sensitively on small variations of the stoichiometry and on the exact preparation conditions, as well as the actual temperature of a sample. This was recently investigated on d-AlNiCo [48].

In some of the d-phases, for example AlNiCo, there is a prominent diffuse layer-line system that doubles the period along the periodic c-axis (Fig. 14). That means there is also disordering with respect to the periodic axes in d-phases. Corresponding to the extended diffuse sheets are, in direct space, 1D structural elements. Apparently there is a gradual change from more or less completely diffuse layers to layers of superstructure ("satellite") reflections in different compounds which indicate lateral short-range correlations between the columns. Generally one observes both types of diffuse contributions. An explanation of this layer-type diffuse scattering is given by 1D columns of icosahedral clusters along the c-axis [9]. The diffuse intensity modulations within the diffuse layers may be discussed in terms of atomic displacements and chemical ordering within the icosahedral clusters. In-situ high-temperature experiments up to 1000°C, i.e. close to the melting point (around 1060°C) showed that diffuse layers vanish upon heating, and recover upon cooling reversibly. A scenario was developed for the structural ordering steps during cooling [10], which includes displacive and migrational short-range ordering processes, involving in-plane displacements and, subsequently, a formation of short-range ordered domains. The availability of modern experimental techniques for recording diffuse scattering contributions, such as area detectors for x-ray and neutron diffraction, even allows a semi-quantitative data evaluation. In the mentioned example of d-AlNiCo a tentative analysis of the integrated diffuse intensities by critical power laws sheds some light on the dimensionality of the individual steps in the complicated order/disorder transition in decagonal quasicrystals [48].

References

1. S. C. Marais, V. Heine, C. M. Nex, and E. K. H. Salje. Phys. Rev. Lett. **66**, 2480 (1991).
2. J. M. Cowley. Diffraction Physics, North-Holland, Amsterdam (1981).
3. H. Boysen. Phase Trans. **55**, 1 (1995).
4. H. Boysen, F. Frey, H. Schrader, G. Eckold. Phys. Chem. Minerals **17**, 629 (1991).
5. F. Frey. Z. Kristallogr. **212**, 257 (1997).
6. F. Frey. 'Diffuse Scattering by Domain Structures'. In: *Local Structure from Diffraction.* ed. by S. J. L. Billinge, M. F. Thorpe (Plenum Press, New York 1998), pp. 295.
7. F. Frey, H. Boysen. Acta Crystallogr. **A37**, 819 (1981).
8. F. Frey. Mat. Sci. Eng. **294-296**, 175 (2000).
9. F. Frey, E. Weidner, K. Hradil, M. de Boissieu, R. Currat, K. Shibata, A. P. Tsai, J. Sato. Phil. Mag. A **80**, 2375 (2000).
10. F. Frey, E. Weidner, K. Hradil, M. de Boissieu, R. Currat, K. Shibata, A.-P. Tsai. *Conference Quasicrystals 2001 at Sendai, Japan sep. 24.-28. 2001*, Contrib. 2I01.
11. A. J. C.Wilson. *X-Ray Optics.* (Methuen, London 1962).
12. H. Jagodzinski, F. Frey. In: *Int. Tables* Vol.B, ed. U. Shmueli (Kluwer, Dordrecht 1993), Ch. 4. 2.
13. H. Jagodzinski. Progr. Crystal Growth and Charact. **114**, 47 (1987).
14. H. Jagodzinski. Acta Crystallogr. **2**,201,208,298 (1949).
15. H. Jagodzinski. Acta Crystallogr. **7**, 17 (1954).
16. H. Jagodzinski, M. Korekawa. Suppl. 3, Geochimica et Cosmochimica Acta **1**, 555 (1972).
17. J. Kakinoki, Y. Komura. J. Phys. Soc. Japan **7**, 30 (1952), **9** 169, 177 (1954).
18. J. Cowley. Acta Crystallogr. **A32**, 83 (1976).
19. K.-H. Jost. *Röntgenbeugung an Kristallen.* (Heyden, Rheine 1975).
20. B. E. Warren. *X-Ray Diffraction.* (Addison-Wesley, Reading 1969).
21. W. Adlhart. Acta Crystallogr. **A37**, 794 (1981).
22. T. R. Welberry, B. D. Butler. Chem. Rev. **95**, 2369 (1995).
23. T. R. Welberry. Rep. Progr. Phys. **48**, 1543 (1985).
24. A. Guinier. X-ray diffraction in crystals, imperfect solids and amorphous bodies. (Freeman, San Francisco 1963).
25. P. Launois, R. Moret. J. Phys. IV France **10**, Pr 3-193 (2000).
26. L. H. Schwartz, J. B. Cohen. *Diffraction from materials.* (Acad. Press, New York 1977), ch. 7.
27. W. Steurer, F. Frey. Phase Transitions. **67**, 319 (1998).
28. R. Hosemann, S. N. Bagchi. *Direct analysis of diffraction by matter.* (North-Holland, Amsterdam 1962).
29. S. B. Hendricks, E. Teller. J. Chem. Phys. **10**, 147 (1942).
30. C. Thompson, M. E. Misenteiner, S. C. Moss. Acta Crystallogr. **44**, 6 (1988).
31. E. Blanc, H. B. Buergi, R. Restori, D. Schwarzenbach, Ph. Ochsenbein. Europhys. Lett. **33**, 205 (1996).
32. A. Babkevich, F. Frey, R. Neder, B. I. Nikolin. Phys. stat. sol. (a) **155**, 3 (1996).

33. R. Lifshitz. *Conference Quasicrystals 2001 at Sendai, Japan sep. 24.–28. 2001*, Contrib. 3IO2.
34. L.-H. Tang. Phys. Rev. Lett. **64**, 2390 (1990).
35. M. V. Jaric. Mat. Sci. Forum **22-24**, 345 (1987).
36. M. V. Jaric, D. R. Nelson. Phys. Rev B **37**, 4458 (1988).
37. M. Widom. Phil Mag. Lett. **64**, 297 (1991).
38. Y. Ishii. Phys. Rev. B **45**, 5228 (1992).
39. Y. Ishii. Mat. Sci. Eng. **294-296**, 377 (2000).
40. M. de Boissieu, M. Boudard, B. Hennion, R. Bellissent, S. Kycia, A. Goldman, C. Janot, M. Audier. Phys. Rev. Lett. **75**, 89 (1995).
41. P. C. Gibbons, K. F. Kelton. Phil. Mag. **B59**, 593 (1993)
42. J. L. Robertson, S. C. Moss. J. of non-cryst. solids **153/154**, 77 (1993).
43. A. I. Goldman, C. A. Guryan, P. W. Stephens, J. M. Parsey, G. Aeppli, H. S. Chen, F. W. Gayle. Phys. Rev. Lett. **61**, 1962 (1988).
44. A. Proult, P. Donnadieu, K. Wang, P. Garoche. J. Phys. I France **5**, 1615 (1995).
45. J.-L. Lei, R. Wang, C. Hu, and D.-H. Ding. Phys. Rev. B **59**,822 (1999).
46. E. Weidner, J.-L. Lei, F. Frey, R. Wang. To be published.
47. E. Weidner, F. Frey, K. Hradil. Phil. Mag. A **81**, 2375 (2001).
48. E. Weidner, F. Frey, K. Hradil, M. de Boissieu, G. McIntyre, A. P. Tsai. *Conference Quasicrystals 2001 at Sendai, Japan sep. 24.-28. 2001*, Contrib. 2012.

Inelastic Neutron Scattering from Structural Excitations

Götz Eckold

Institut für Physikalische Chemie, Universität Göttingen, Germany

Abstract. In this lecture, a short overview is given over the characteristic features of inelastic neutron scattering along with some selected applications for the investigation of dynamic processes in condensed matter. Of course, a comprehensive treatment of the entire field of neutron scattering cannot be provided within a single lecture. Hence, I restricted myself on the treatment of structural excitations thereby skipping the important field of magnetic excitations. After some introductory remarks about the role of neutrons in condensed matter research, a brief summary of scattering processes and correlations will be given along with a short review on experimental techniques. The potential of inelastic neutron scattering will be demonstrated with the help of a number of different examples dealing with collective excitations, diffusion processes, tunnelling, polymer dynamics, and kinetic studies. More detailed information about individual aspects of different applications and techniques may be found in the series of lecture notes of summer schools organised by FZ Jülich or PSI, Villigen [1].

1 Why Neutrons?

It is well known that the microscopic structure of condensed matter can be investigated by diffraction methods using radiation with wavelengths comparable to the inter-atomic spacing of the order Å. Similarly, spectroscopic techniques allow to determine dynamical properties if the characteristic frequency of the radiation is adapted to the problem under consideration. In condensed matter research, the interesting frequency regime of structural excitations extends from THz to MHz, roughly speaking. For electromagnetic radiation, the frequency ν and the wavelength λ are related via the velocity of light:

$$\nu = \frac{c}{\lambda} \quad \text{for electromagnetic radiation.} \tag{1}$$

Hence, X-rays with wavelengths in the Å-region are suitable for the investigation of atomic or molecular structures. At the same time, this radiation exhibits frequencies of 10^{18} Hz far above the frequency regime of structural excitations. Even if the event of modern synchrotron sources and advanced techniques for monochromatisation meanwhile allow spectroscopic studies in the THz-range with X-rays, it is infrared or microwave spectroscopy with frequencies between THz and MHz which directly probes the time-scale of dynamic processes. This radiation is, however, unable to yield information about spatial correlations due to its long wavelength.

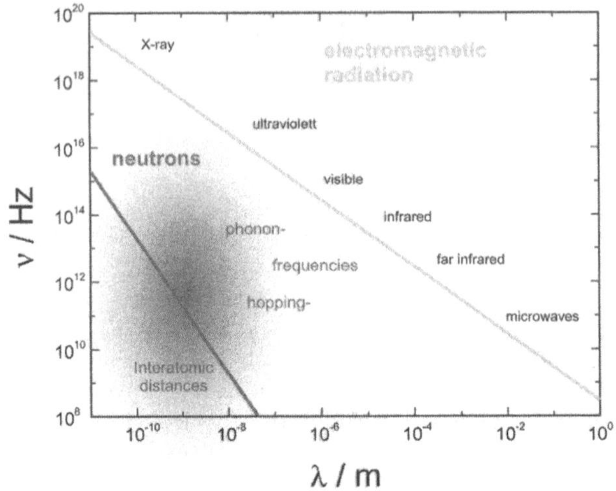

Fig. 1. Frequency-wavelength diagram for electromagnetic radiation and neutrons

Neutrons are unique for the investigation of atomic motions in condensed matter since they exhibit both, frequencies of the order THz and wavelengths of the order *Å*, according to de Broglie's relation

$$E = \frac{m}{2}v^2 = \frac{h^2}{2m\lambda^2} = h\nu\,. \tag{2}$$

In Fig. 1 the frequency-wavelength relationship is shown for neutrons and electromagnetic radiation, respectively, along with the interesting regime of structural excitations. Obviously, neutrons are ideally suited to resolve dynamical processes, both, in time and in space.

Another fundamental difference between neutrons and electromagnetic radiation is its type of interaction with matter. Having no charge, neutrons are able to penetrate even large samples. There is no Coulomb-interaction with the electrons; rather, neutrons are scattered by the nuclei. Moreover, neutrons carry a spin and a magnetic moment. Consequently, there is a magnetic interaction with the nuclear or electron spins in the sample. This gives rise to magnetic scattering which will, however, not be considered in the present paper. The third kind of interaction of neutrons with matter is true absorption by nuclear reactions such as (n,α)- or (n,γ)-processes. This is used for the shielding and also the detection of neutrons.

2 Scattering and Correlation

Let us concentrate on nuclear scattering. Since nuclei exhibit typical diameters of the order 10^{-15}m which is much smaller than the neutron wavelength, scattering from a single nucleus is almost isotropic.

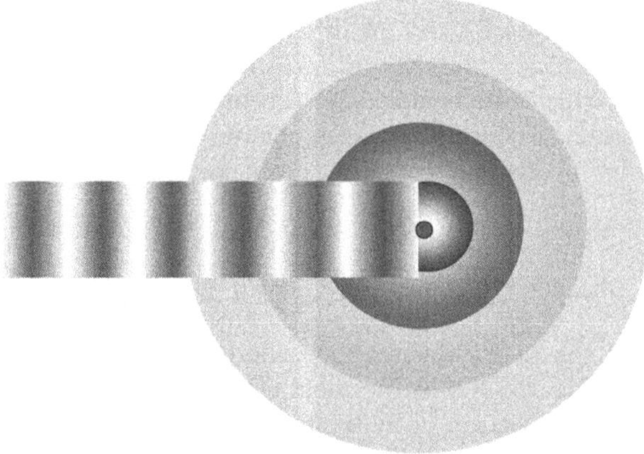

According to Huygens' principle the scattered wave may therefore be described by a simple spherical wave with a characteristic amplitude described by a scattering length b:

$$\text{scattered wave} = -\frac{b}{|\mathbf{r}|}\exp[i\mathbf{k}\,(\mathbf{r}-\mathbf{R}_\circ)]\,. \tag{3}$$

The interaction potential between the nucleus located at $\mathbf{R}_\circ$ and the neutron at $\mathbf{r}$ may be described by a δ-function:

$$V(\mathbf{r}) = b\cdot\frac{h^2}{2\pi m_n}\delta(\mathbf{r}-\mathbf{R}_\circ) \tag{4}$$

(where m_n is the neutron's mass). It is called the *Fermi pseudo-potential.*

The scattering lengths vary from element to element in a non-systematic way and even from isotope to isotope of the same element. Moreover, the relative orientation of nuclear spins with respect to the neutron spin affects the overall scattering length. Hence, not all chemically identical atoms exhibit the same scattering power. It is therefore not sufficient to determine a mean scattering length but also its variance.

The scattering from an arrangement of atoms like a solid is just the superposition of all partial waves emanating from the individual nuclei. The scattered intensity would be the result of interference effects as long as all chemically identical atoms behave identical. The interference pattern, the coherent scattering, will therefore be governed by the average scattering lengths which are called the coherent scattering lengths b_{coh}

$$b_{coh} = \langle b\rangle\,. \tag{5}$$

These are related to the coherent scattering cross sections σ_{coh} via

$$\sigma_{coh} = 4\pi\langle b\rangle^2\,. \tag{6}$$

In addition, there is another contribution to the scattered intensity which is not affected by the interference of different partial waves as long as isotope distribution and nuclear spin orientation is at random and unpredictable. This part of the scattering is determined by the properties of individual atoms and is called incoherent scattering. It is associated with the incoherent scattering length b_{inc} or the incoherent scattering cross section σ_{inc}:

$$\begin{aligned} b_{inc} &= \sqrt{\langle b^2\rangle - \langle b\rangle^2}\,, \\ \sigma_{inc} &= 4\pi b_{inc}^2 = 4\pi\Big(\langle b^2\rangle - \langle b\rangle^2\Big) = \sigma_{tot} - \sigma_{coh}\,. \end{aligned} \tag{7}$$

For a number of selected elements the coherent and incoherent cross sections are illustrated in Fig. 2. One of the most striking features is the extraordinary large incoherent cross section of hydrogen (^{1}H). The scattering from hydrogen-containing samples will therefore, in most cases, be dominated by incoherent scattering which reflects the single-particle behaviour of protons. Hence, the dynamics of polymers, the dynamics of hydrogen in biological systems, the diffusion of hydrogen in metals or hydrocarbons in zeolites, e.g., are phenomena which can be well studied by neutrons. On the other hand, neighbouring atoms like Fe and Cr, e.g., which are hardly distinguishable by X-ray scattering provide a well detectable scattering contrast if neutrons are used.

In a typical scattering experiment, a well defined neutron beam which is characterised by a wave vector $\mathbf{k}_i$ ($|\mathbf{k}_i| = \frac{2\pi}{\lambda_i}$) hits the sample.

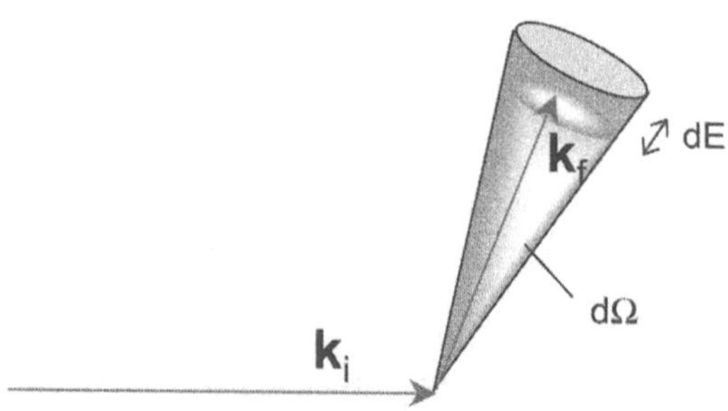

The scattered neutrons may be described by the wave vector $\mathbf{k}_f$ which differs from $\mathbf{k}_i$, both, in direction and in modulus. Remembering that momentum and energy are given by $\hbar\mathbf{k}$ and $\hbar^2\mathbf{k}^2/2m_n$, respectively, this scattering process is characterised by the momentum transfer

$$\hbar\mathbf{Q} = \hbar(\mathbf{k}_i - \mathbf{k}_f) \tag{8}$$

and the energy transfer

$$E = \hbar\omega = \frac{\hbar^2}{2m_n}(\mathbf{k}_i^2 - \mathbf{k}_f^2) \tag{9}$$

from the neutron to the sample.

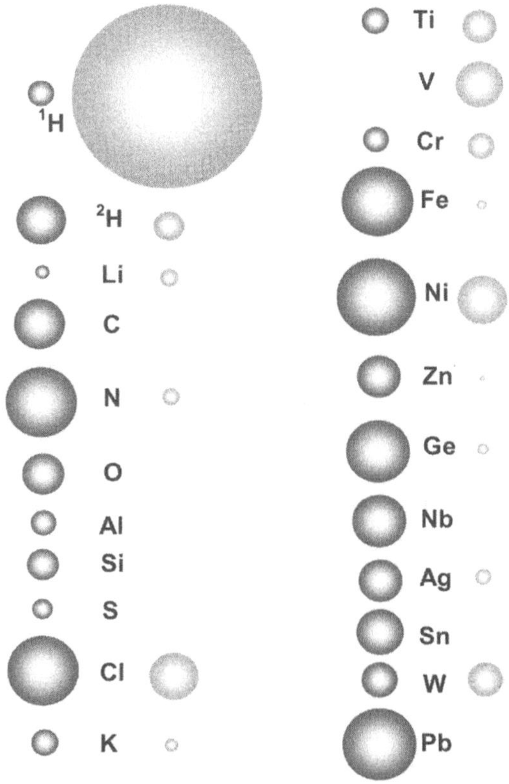

Fig. 2. Some selected coherent (dark) and incoherent (light) scattering cross sections

A detector collects all those neutrons which are scattered into a solid angle element $d\Omega$ around $\mathbf{k}_f$ with energies between E_f and $E_f + dE$ and the intensity is determined by the double differential cross section $d^2\sigma/d\Omega dE$. If, during the scattering process, the sample changes its quantum state from an initial state $\rho_\circ$ to a final state ρ, this cross section can be calculated using Fermi's Golden Rule since the interaction with the neutron according to (4) can be regarded as a small perturbation:

$$\frac{d^2\sigma}{d\Omega dE} = \frac{|\mathbf{k}_f|}{|\mathbf{k}_i|}\left(\frac{2\pi m_n}{h^2}\right)^2 \sum_{\rho_\circ,\rho} p_{\rho_\circ} |\langle \mathbf{k}_f \rho | V | \mathbf{k}_i \rho_\circ \rangle|^2 \delta(E + E_{\rho_\circ} - E_\rho). \qquad (10)$$

Here, $p_{\rho_\circ}$ denotes the thermal population factor of the initial state. Since the state of the sample itself cannot be analysed, the summation about all possible initial and final states has to be carried out. The δ-function guarantees the energy conservation. Details of the derivation are beyond the scope of this lecture and may be found elsewhere [2]. The sum on the right hand side of (10) does only depend on the momentum transfer or the scattering vector $\mathbf{Q}$ and the energy transfer or the frequency ω. It contains all the information about the dynamical behaviour of the sample. Hence, we can introduce the so-called scattering-function $S(\mathbf{Q},\omega)$ to write shortly

$$\frac{d^2\sigma}{d\Omega dE} = \frac{|\mathbf{k}_f|}{|\mathbf{k}_i|} S(\mathbf{Q},\omega)\,. \tag{11}$$

Since only the incoherent and coherent averages of the individual scattering lengths are known the scattering-function itself is the sum of two contributions:

$$S(\mathbf{Q},\omega) = \frac{1}{4\pi}\sum_i \sigma_{inc}^{(i)} S_{inc}^{(i)}(\mathbf{Q},\omega) + \frac{1}{4\pi}\sum_i\sum_k \sqrt{\sigma_i}\sqrt{\sigma_k}^* \cdot S_{coh}^{(i,k)}(\mathbf{Q},\omega), \tag{12}$$

where $S_{inc}^{(i)}(\mathbf{Q},\omega)$ is the incoherent scattering function for particles of type (i) and $S_{coh}^{(i,k)}(\mathbf{Q},\omega)$ is the coherent scattering function which describes the interference pattern resulting from particles of type (i) and (k).

After some mathematics, it can be shown [2] that $S_{inc}^{(i)}(\mathbf{Q},\omega)$ is just the Fourier-transform of the space-time self-correlation function $G_s^{(i)}(\mathbf{r},t)$ of particles (i) which is the probability of finding a particle after a time t at the position $\mathbf{r}$ if it was at the origin at $t=0$:

$$G_s^{(i)}(\mathbf{r},t) = \frac{1}{N}\sum_j \int d\mathbf{r}' \Big\langle \delta\big(\mathbf{r}' - \mathbf{R}_j(0)\big)\delta\big(\mathbf{r}' + \mathbf{r} - \mathbf{R}_j(t)\big)\Big\rangle_T\,. \tag{13}$$

This function includes all information about the single particle dynamics.

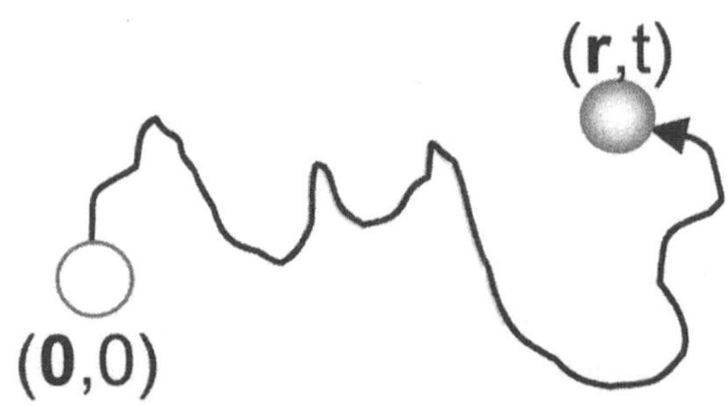

Similarly, $S_{coh}^{(i,k)}(\mathbf{Q},\omega)$ is the Fourier-transform of the space-time pair correlation function $G^{(i,k)}(\mathbf{r},t)$. This so-called van-Hove correlation function describes the probability of finding a particle of type (k) after a time t at the position $\mathbf{r}$ if a particle of type (i) was at the origin at $t=0$:

$$G^{(i,k)}(\mathbf{r},t) = \frac{1}{N}\sum_{j,j'} \int d\mathbf{r}' \Big\langle \delta\big(\mathbf{r}' - \mathbf{R}_j(0)\big)\delta\big(\mathbf{r}' + \mathbf{r} - \mathbf{R}_{j'}(t)\big)\Big\rangle_T\,. \tag{14}$$

$G^{(i,k)}(\mathbf{r},t)$ reflects the collective dynamics of the scattering system.

Obviously, neutron scattering provides detailed microscopic information about static and dynamics properties. In the following, we will discuss some of the experimental techniques which are suitable for the investigation of different kinds of structural excitations.

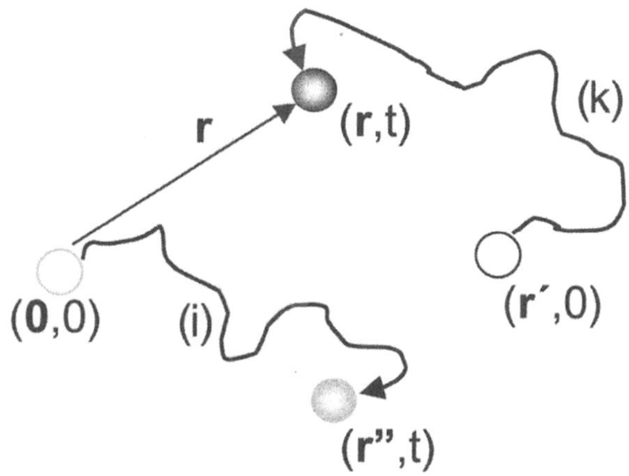

3 Techniques

Free neutrons can be produced by nuclear reactions in fission reactors or in spallation sources. They exhibit energies in the MeV-region and have to be slowed down to thermal energies in a moderator before they can be used for scattering experiments. The Maxwellian energy spectrum of the neutrons can be shifted by adapting the temperature of the moderator. While thermal sources yield a maximum neutron flux at a wavelength of about 1.6Å, cold (hot) moderators operated at temperatures near 25 K (2000 K) shift this maximum to 5.5 (0.6) Å. The choice of cold, thermal or hot neutrons as well as a suite of different experimental techniques allows to cover a range of almost nine orders of magnitude in energy-transfer (from 10^{-9} eV up to 1 eV) and four orders of magnitude in momentum transfer (from $< 0.1\,Å^{-1}$ up to some $100\,Å^{-1}$) as shown in Fig. 3.

The different spectrometer types are essentially characterised by their method of monochromatisation of both, the primary and the scattered beam. One can distinguish between filter techniques, time-of-flight and crystal monochromators.

For large energy and momentum transfers beryllium-filter spectrometers are suitable which use hot neutrons. Energies in the meV-range are covered by time-of-flight (TOF) and three-axes (TAS) instruments, while the μeV-range is the domain of back-scattering spectrometers. Even lower energies down to neV can be accessed by spin-echo techniques which use spin-polarised neutrons.

In the following section, different spectrometer types will be described along with typical examples for their respective application.

4 Applications

A sketch of some dynamical phenomena which can be investigated by inelastic neutron scattering is shown in Fig. 4. Atomic vibrations, phonons (and also magnons) are typical structural excitations within the THz-frequency regime.

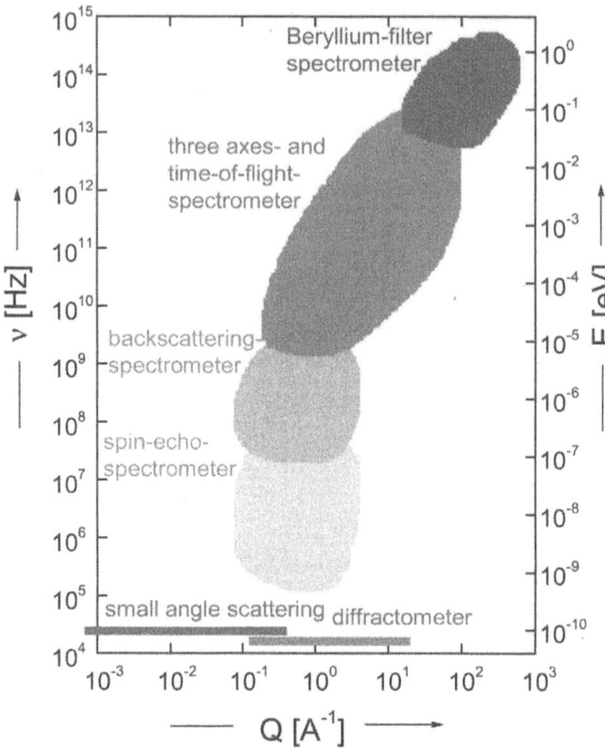

Fig. 3. Accessible ranges of energy- and momentum transfer for different types of spectrometers (for comparison the Q-ranges of small-angle scattering instruments and diffractometers without energy resolution are also included)

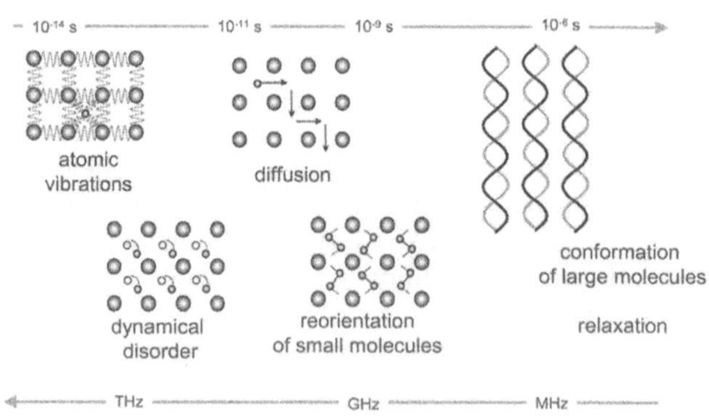

Fig. 4. Some dynamical phenomena in condensed matter

Dynamical disorder associated with jumps of atoms between different accessible sites occur quite frequently on a time scale between ps and ns. The elementary steps of diffusion processes are strongly temperature dependent and can be probed with neutrons if jump frequencies between some 100 GHz and some 10 MHz are involved. The reorientation of small molecules like N_2 or CH_4 in solids affects the dynamical behaviour of molecular crystals in the frequency range around GHz. On a time scale of *ns* to *μs* even the conformational dynamics of large molecules like polymers may be studied in soft matter.

4.1 Collective Excitations, Phonons – The THz-Regime

Lattice vibrations or phonons are described by their wave vector **q**, their energy, and their polarisation vector, i.e., the displacement pattern of all individual atoms. Since the phonon energy depends on modulus of **q** as well as on its direction, the determination of the dispersion requires the orientation of single crystals with respect to the neutron beam. The most appropriate instrument for the determination of the phonon dispersion and eventually also of the polarisation vector is a three-axes spectrometer as shown schematically in Fig. 5.

The white neutron beam is extracted from a reactor source through a beam tube inserted into the biological shield. By Bragg-reflection from a single crystalline monochromator typically made of pyrolytic graphite, copper, germanium or silicon, a narrow energy band is selected. The energy can easily be changed by a variation of the Bragg angle. The monochromatised beam hits the sample which is precisely oriented on a sample table. Due to inelastic scattering, the scattered beam will in general be again polychromatic. Another single crystal acting as an analyser selects neutrons of a particular energy, which are finally registered in the detector. Monochromator, sample, and analyser define the three axes which define the incident energy, the scattering angle, and the final energy, respectively. All these variables can be varied independently, thus providing a unique flexibility of this spectrometer type. While Fig. 5 shows the layout of

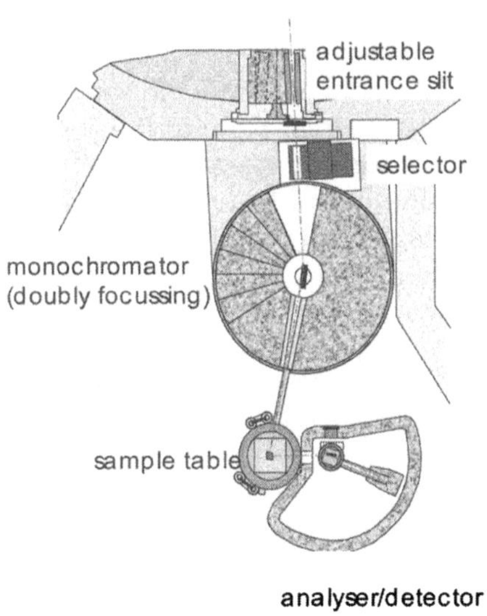

Fig. 5. Layout of the three-axes spectrometer PUMA (FRM-II)

Fig. 6. The three-axes spectrometer UNIDAS (FRJ-2)

the PUMA-spectrometer which is currently being installed at the new Munich reactor source, Fig. 6 gives an impression of the existing three-axes spectrometer UNIDAS at the FRJ-2 reactor in Jülich. Unlike optical spectroscopy which is sensitive to lattice vibrations with very long wavelength only, neutrons can principally detect phonons with arbitrary wave vectors. The whole variety of phonon branches reflects the details of interatomic interactions. As an impressive example, the phonon dispersion of the high-temperature superconductor Nd_2CuO_4 is displayed in Fig. 7 for phonon wave vectors along the main symmetry directions. Since the mechanism for the superconductivity in ceramics is still under debate, the knowledge of the phonon spectrum and of possible anomalies due to electron-phonon coupling is of particular interest. The 21 phonon branches are almost completely determined by Pintschovius et al. [3]. Most of the features could be explained by phenomenological model calculations based on pair interactions between the atoms. In conventional superconductors, an anomalous phonon softening was observed either close to $\mathbf{q} = 0$ as in Nb or well inside the Brillouin zone as in TaC. In high-temperature superconductors no such effects have been detected and, hence, it is believed that electron-phonon coupling is not the essential mechanism for superconductivity in these ceramic materials.

If, as a function of external parameters like temperature or pressure, some particular phonon mode becomes soft, a lattice instability may result. The study of those soft-mode phase transitions is an actual field of research.

In a recent study, the dynamical behaviour of uranium has been investigated. It is well known, that in orthorhombic α-uranium a charge density wave (CDW) is formed at temperatures below $T_\circ \sim 43K$ which is associated with a rather complicated displacement pattern of the individual atoms. Hence, the ordering processes in the electron subsystem and in the atomic structure are correlated. It could be shown that the CDW is characterised by a wave vector $\mathbf{q}_{CDW} = \pm 0.490$ $a^* \pm 0.131\ b^* \pm 0.225 c^*$ which is incommensurate with the underlying lattice, i.e. it yields a modulated structure rather than a simple superstructure.

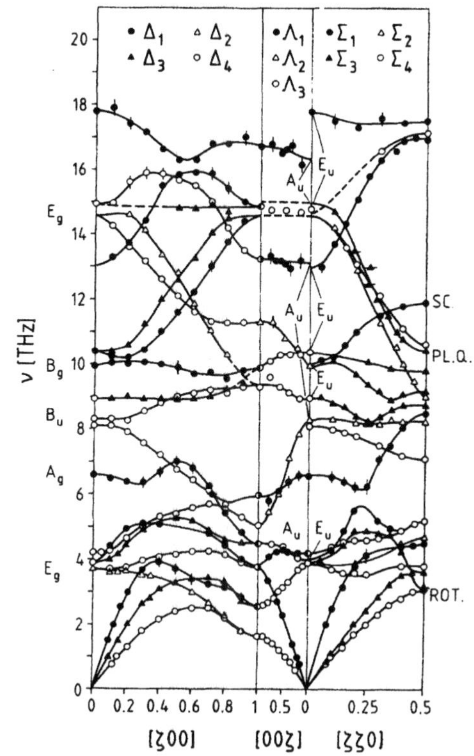

Fig. 7. Phonon dispersion of Nd_2CuO_4. After [3]

It was usually assumed that the phase transition is driven by a phonon instability and a strong electron-phonon coupling and elastic strain fields are present in this system. Different theoretical models predict, however, different points in **q**-space where the soft mode should actually condensate and becomes a static modulation of the lattice.

Using inelastic neutron scattering, Marmeggi et al. [4] investigated the phonon dispersion in the vicinity of $\mathbf{q}_{CDW}$.

In Fig. 8 the results of two different energy scans taken at $55\,K$ for different scattering vectors **Q** within the (101)-Brillouin zone are shown. In the upper part ($\mathbf{Q} = (1.5,\ 0,\ 1)$) the phonon wave-vector is just $0.5a^*$ and two phonon peaks on the energy-gain and the energy-loss side are clearly resolved at about $\pm 2meV$. If, however, the wave vector is tilted from the a^*-axis to $\mathbf{q} = (0.5,\ 0.127,\ 0.205)$ the phonon energy is even lower. Hence, it is not the high-symmetry point where the phonon frequency is lowest.

In a 3D-representation (Fig. 9) the energy is plotted as a function of two components of the wave vector, along and perpendicular to the a^*-axis. It is clearly seen that this dispersion surface exhibits two well defined minima of the phonon energy for wave vectors well separated from the orthorhombic a^*-axis.

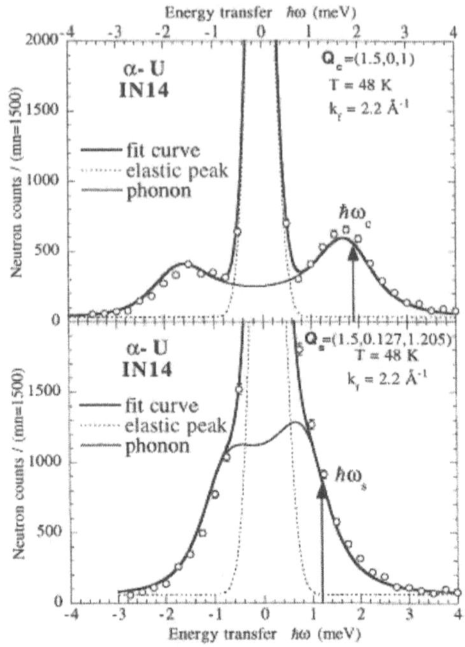

Fig. 8. Energy spectra of α-uranium taken at T=55 K for two phonon wave vectors, **q**=(0.5,0,0) (top) and **q**=(0.5,0.127,0.205) (bottom) within the Brillouin zone (101). After [4]

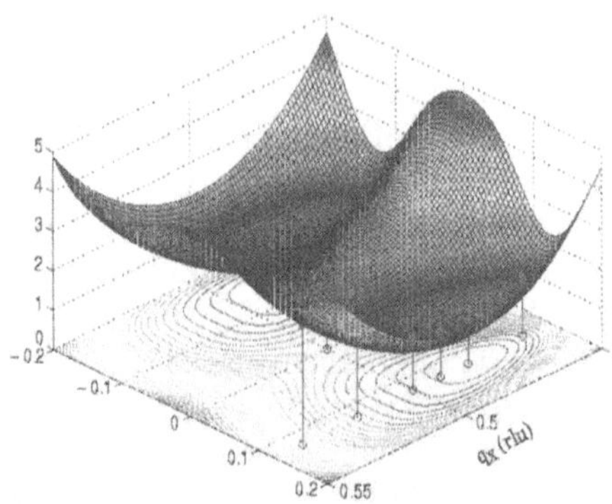

Fig. 9. 3D-representation of dispersion surface of α-uranium at T=55 K. After [4]

As a function of temperature, the phonon energy changes drastically. Figure 10 displays the temperature dependence for three values of the phonon wave vectors. Obviously, just at $T_\circ$ there is a condensation of a phonon with a wave vector which agrees remarkably well with $\mathbf{q}_{CDW}$.

This result was taken as a proof for the fact that the CDW-transition is associated with an electron-phonon driven lattice instability. Moreover, it gives

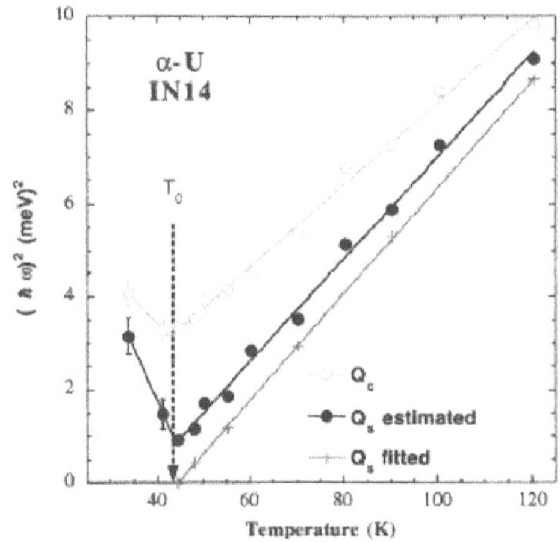

Fig. 10. Temperature dependence of the square of the phonon energy for three different wave vectors, $\mathbf{q}$=(0.5,0,0) (top), $\mathbf{q}$=(0.5,0.125,0.22) (middle), $\mathbf{q}$=(0.485,0.13,0.21) (bottom, fitted minimum of the dispersion surface as shown in Fig. 9) After [4]

evidence that Fermi-surface nesting effects play an important role in this system and determine the wave vector $\mathbf{q}_{CDW}$.

It should be noted, that the temperature dependence of the phonon frequencies can well be described in the framework of the classical Landau theory: As shown in Fig. 10, the squared energy is proportional to $|T - T_\circ|$ over quite a large range of temperatures.

Another example for the unique possibilities of neutron scattering is the determination of the origin of the negative thermal expansion observed in some unusual materials.

Thermodynamics reveals that the coefficient of thermal expansion β is related to the volume dependence of phonon frequencies. For cubic materials the following relation holds:

$$\beta = \frac{3}{V(c_{11} + 2c_{12})} \sum_{\mathbf{q},j} \tilde{C}_{\mathbf{q},j}\, \gamma_{\mathbf{q},j} \,, \tag{15}$$

where c_{11} and c_{12} are two elastic constants, V is the molar volume and $\tilde{C}_{\mathbf{q},j}$ are the contributions of the individual phonon states $(\mathbf{q}, j)$ to the molar heat capacity. $\gamma_{\mathbf{q},j}$ is the so-called *mode-Grüneisen parameter*, defined as the logarithmic derivative

$$\gamma_{\mathbf{q},j} = -\frac{\partial \ln \omega_{\mathbf{q},j}}{\partial \ln V} \,. \tag{16}$$

The summation runs over all phonons as characterised by their wave vector $\mathbf{q}$ and the branch index j ($j = 1, ..., 3N$, N number of atom per primitive unit cell). Usually, phonon frequencies will increase if the volume is reduced due to the enhanced interaction strength thus leading to positive mode-Grüneisen

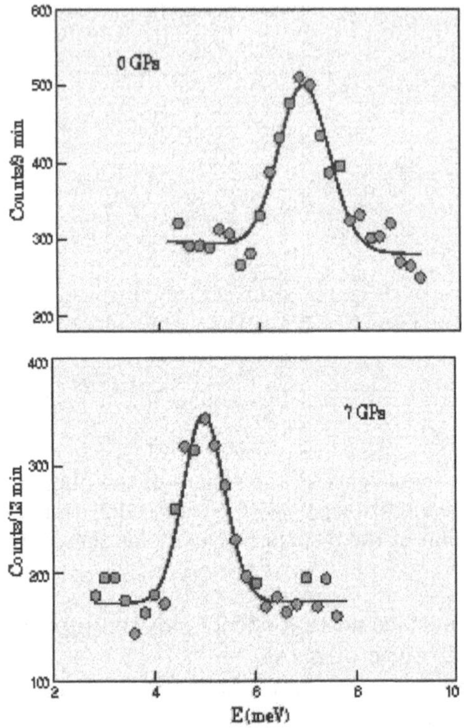

Fig. 11. Zone-boundary acoustic phonon in GaSb at two different pressures. After [5]

parameters. If, however, a material exhibits a negative coefficient of thermal expansion, (13) demands that there has to be a considerable number of phonons which softens on reducing the volume. Hence, the origin of negative thermal expansion is dynamical in nature and the determination of those phonons which yield negative Grüneisen parameters is a task for inelastic neutron scattering. An elucidating example is provided by the system GaSb which has been investigated by Klotz et al. [5]. These authors have determined the frequencies of acoustic phonons as a function of pressure and, hence, as a function of volume. Figure 11 shows the energy spectrum of a zone-boundary phonon at two different pressures. Obviously, there is an unusual strong softening from about $7\,meV$ at ambient pressure to $5\,meV$ at $7\,GPa$ leading to a negative Grüneisen parameter.

These experiments were repeated for a considerable number of phonons with different wave vectors and the resulting dispersion of the mode-Grüneisen parameters is displayed for wave vectors in the main symmetry directions in Fig. 12. It is easily seen that extraordinary negative values of γ are found for wave vectors close to the zone boundary, while γ is small and positive near the zone center (Γ-point, $\mathbf{q} = 0$). This is the reason why in ultrasonic experiments which yield the slope of acoustic phonons for very long wave lengths no anomalies could be detected. Hence, the neutron data prove that the short wave length phonons are responsible for the unusual thermal behaviour of GaSb.

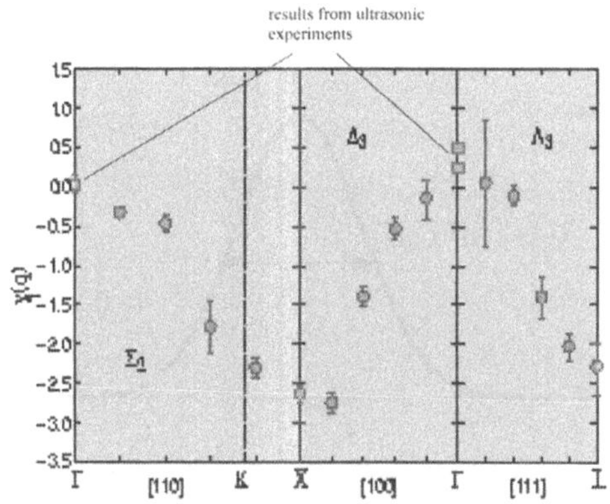

Fig. 12. Dispersion of mode-Grüneisen parameters for acoustic phonon in GaSb. After [5]

4.2 Diffusion – Between GHz and THz

Not only collective excitations can be probed with neutrons but also single particle motion like diffusion. In continuous systems it is governed by the well known Fick's law for the self-correlation function $G_s(\mathbf{r},t)$:

$$\frac{\partial G_s(\mathbf{r},t)}{\partial t} = D\Delta G_s(\mathbf{r},t) \tag{17}$$

which has the solution:

$$G_s(\mathbf{r},t) = \frac{1}{\sqrt{4\pi Dt}^3} \exp\left[-\frac{\mathbf{r}^2}{4Dt}\right] \tag{18}$$

with the appropriate boundary condition

$$G_s(\mathbf{r},0) = \delta(\mathbf{r})\,. \tag{19}$$

The Fourier-transform in space and time yields the incoherent scattering function

$$S_{inc}(\mathbf{Q},\omega) = \frac{1}{\pi}\frac{D\mathbf{Q}^2}{\omega^2 + (D\mathbf{Q}^2)^2}\,, \tag{20}$$

which describes a quasielastic frequency spectrum, a Lorentzian centered at $\omega = 0$. The line width $\Delta\omega$w (FWHM) increases proportional to the square of the scattering vector. From the plot $\Delta\omega\mathbf{Q}^2$ the diffusion coefficient D can be easily determined as the slope of the so-called diffusion-line.

Quite frequently, diffusion processes which exhibit a weak Q-dependence only, are studied by time-of-flight techniques. Thermal and cold neutrons have velocities in the range between some $100\,ms^{-1}$ and several $1000\,ms^{-1}$. If a pulsed (chopped) neutron beam is used, neutrons of different energies can therefore be

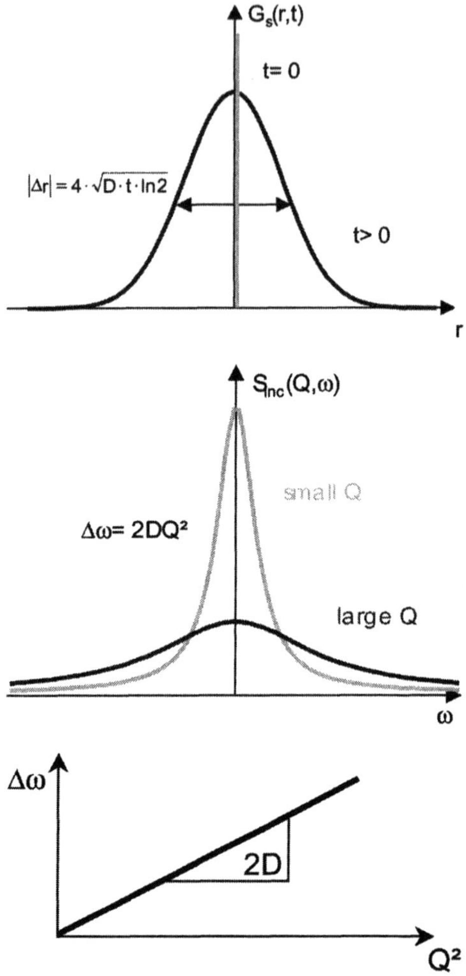

well separated by the time needed to travel over distances of several meters, e.g., from a chopper to the detector. Figure 13 shows the layout of the IN5-spectrometer at the Institut-Laue-Langevin in Grenoble as well as a photograph of the detector area. The incident neutron beam passes a sequence of four choppers which exhibit a well defined phase shift with respect to each other. Hence, only neutrons of a single energy are able to pass all these choppers and hit the sample. Obviously, most of the neutrons are being absorbed within the choppers and the duty cycle is below 1%.

In order to overcome this disadvantage neutrons which are scattered at different angles and with different energy transfers are selected simultaneously: A range of scattering angles of almost 180° is covered be individual detectors, each equipped with a time-resolution system which allows to determine the time of arrival and, hence, the energy for each detected neutron.

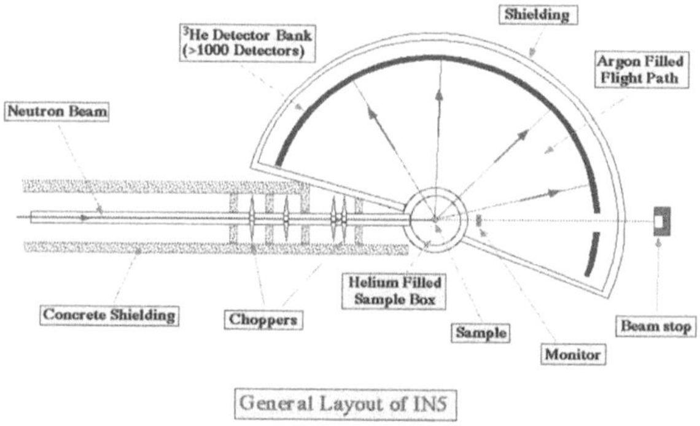

Fig. 13. Layout of the TOF-spectrometer IN5 at the ILL, Grenoble (top) and view of the detector area (below)

While at the IN5 spectrometer the energy of both, the incident and the scattered neutrons are determined by time of flight, there are other techniques which use crystal monochromators for selecting the incident neutrons. Time-of-flight techniques are particularly useful if the neutron source itself provides a time-structure just like pulsed spallation sources.

Hydrogen containing samples are particularly suited for the study of diffusion processes due to the large incoherent cross section of protons. A detailed study of the diffusion of water molecules has been performed by *Texeira et al.* [6]. In fact, the quasielastic scattering was observed which could be fitted to Lorentzians according to (20). For different temperatures from room temperature, $20°C$, down into the supercooled state at $-20°C$, the quasielastic line width is shown as a function of $\mathbf{Q}^2$ in Fig. 14. At very small $\mathbf{Q}$, the line width is proportional to $\mathbf{Q}^2$ as predicted for continuous diffusion. With increasing $\mathbf{Q}$, however, deviations from the straight-line behaviour are readily seen. This effect is more pronounced at lower temperatures where the influence of the ordered structure becomes more and more important and the character of the molecular motion changes towards a jump diffusion with well defined residence times of the order 10 ps.

At $\mathbf{Q} > 1Å^{-1}$ neutrons probe the behaviour of the system at length scales below several Å and, hence, it is no longer allowed to apply the simple model

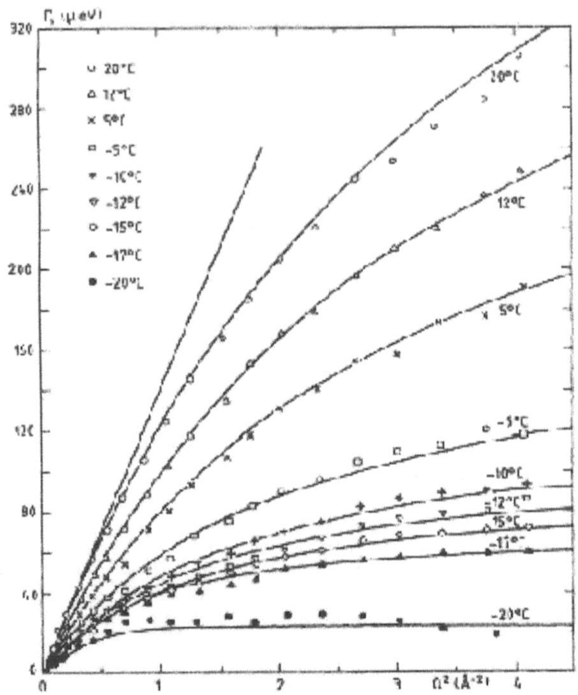

Fig. 14. Q-dependence of the quasielastic line width of water at different teperatures between -20 C° (bottom) and 20 C° (top). After [6]

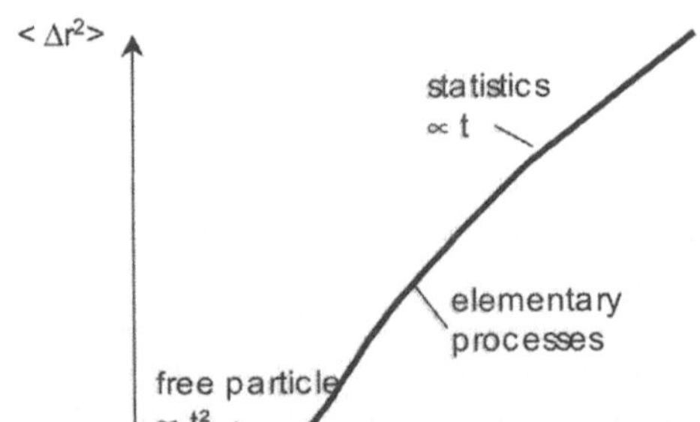

Fig. 15. Time-dependence of the mean-square displacement of diffusing particles, schematically

for continuous diffusion. Rather, the elementary step-like jumps between well defined sites show up. Schematically, the time dependence of the mean-square displacement of a single diffusing particle may be represented by Fig. 15.

At very small times, it behaves like a free particle with constant velocity. At very large times on the other hand, all interactions with other particles can be represented by an effective frictional force and the diffusion obeys the statistical

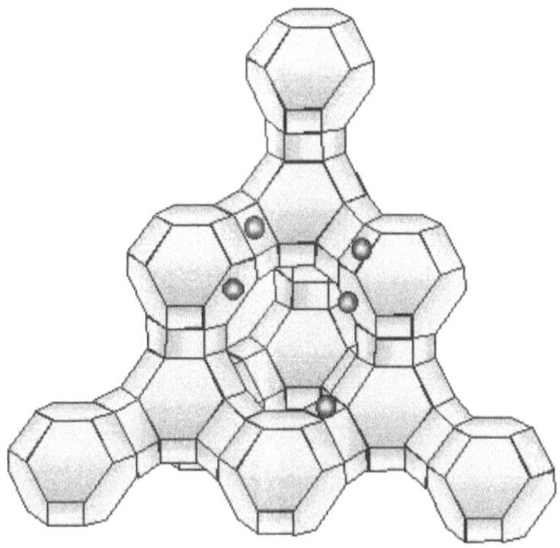

Fig. 16. The faujasite-structure of NaX-zeolite with possible sites for para-xylene molecules, $C_6H_4(CH_3)_2$

laws. The intermediate time regime, however, is the most interesting one, since it provides information about the microscopic nature of single diffusion steps. It is just this regime which is accessible by quasielastic neutron scattering.

In solid systems, detailed information about the residence sites and the residence times of diffusing particles, the corresponding jump frequencies and the jump vectors can be obtained.

A recent example is the determination of pinning sites for organic molecules in zeolites which provide technically important systems as molecular sieves or as materials for heterogeneous catalysis. Frequently, zeolites are alumosilicates which form a network of interconnected cages. In these cages, guest molecules can be imbedded which might diffuse from one cage to the other. The size of the cages as well as the size of the interconnecting channels determine the diffusivity.

Jobic and Bée [7] studied the diffusion process of para-xylene, $C_6H_4(CH_3)_2$, in NaX-zeolite of the faujasite-type which forms particularly large cages as shown in Fig. 16. The xylene guest molecules may be pinned at different sites within these large cages and, from time to time, they jump from one site to another thereby travelling from one cage to another. Neutron experiments yield a **Q**-dependence of the quasielastic line width which is displayed in Fig. 17.

Obviously, there is no similarity with the behaviour expected for the simple continuous diffusion, except at very low **Q**. Rather, the non-monotonic shape directly reflects the discontinuous aspects of diffusion in solids. Using a microscopic model which takes into account discrete jump vectors, it was in fact possible to reproduce the experimental data quantitatively. Moreover, from the jump distances the authors were able to derive the residence sites within the cages of the zeolite structure. From the asymptotic value of the quasielastic line width for large **Q** the characteristic jump frequency can directly be determined as about 0.5 GHz.

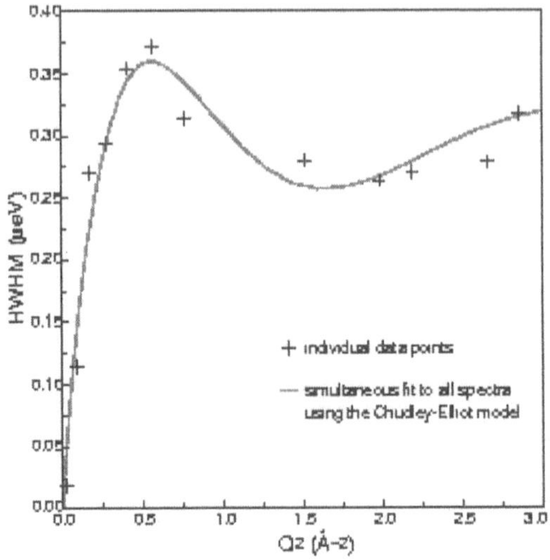

Fig. 17. Q-dependence of the quasielastic line width of the system para-xylene in NaX-zeolite (after [7]). The continuous line is a result of model calculations of jump-diffusion.

4.3 Molecular Rotations, Tunnelling – The GHz-Regime

The rotation of molecular groups in solids is another actual field of research with neutrons. The characteristic energies for the excitation of rotational motions are found in the range between μeV and meV. The most appropriate instrument for their investigation is the so-called *back-scattering spectrometer*. Figure 18 shows a sketch of the IN16 spectrometer at the ILL in Grenoble.

In principle, the back-scattering spectrometer is just a three-axes instrument in which the crystal monochromator as well as the analyser are operated in back-scattering geometry thus providing an extremely good energy resolution even if the divergence of the neutron beam is quite large. As illustrated in Fig. 18, the incident neutron beam is extracted from a neutron guide by two crystal deflectors. This beam then hits a monochromator crystal under normal incidence which selects a narrow wavelength band. Since the monochromator is periodically moving back and forth, the energy of the reflected neutrons is modulated due to the Doppler shift. Hence, as a function of time, neutrons with slightly different energies arrive at the sample. The scattered neutrons are collected and back-reflected by huge analyser crystals which are shown in the photograph. These analysers define a very narrow energy interval of those neutrons which are finally registered in the detectors located close to the sample position. Thus, as a function of time, neutrons which exhibit different energy transfers are counted and accumulated over several periods of the Doppler-drive.

As a particularly elucidating and meanwhile already classical example for the power of back-scattering spectroscopy, a study of molecular rotation in solid methane is chosen. There are several solid phases of methane which differ in the

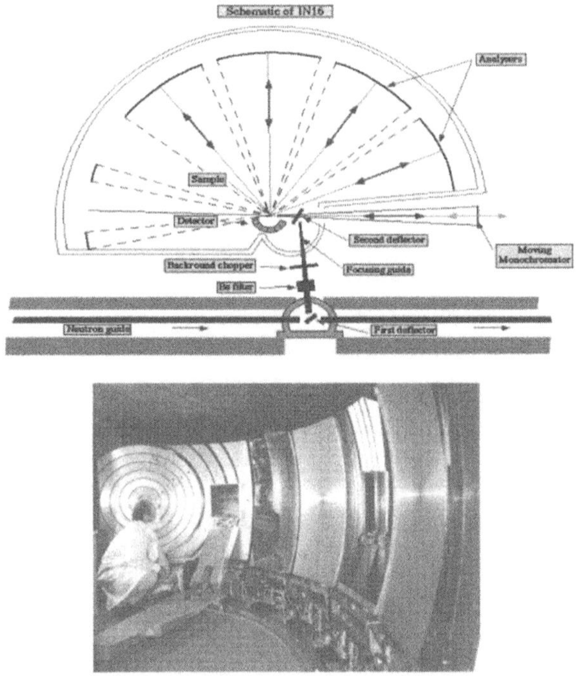

Fig. 18. Layout of the back-scattering spectrometer IN16 at the ILL, Grenoble (top) and view of the analyser array (below)

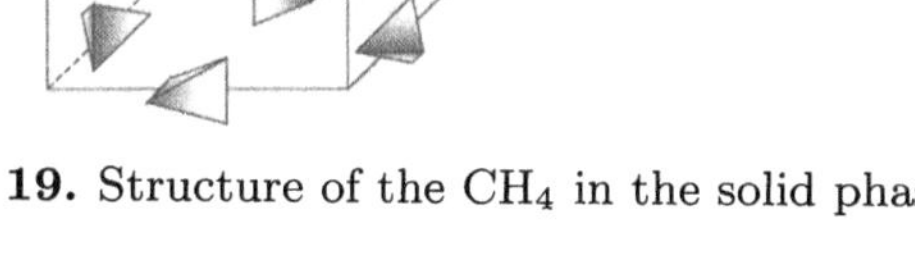

Fig. 19. Structure of the CH_4 in the solid phase II

orientational order of the CH_4-molecules. In the so-called phase II, the tetrahedral CH_4-groups are arranged on a fcc-lattice as shown in Fig. 19.

While three of these molecules are orientationally ordered, the fourth does not exhibit any preferred orientation and is therefore dynamically disordered. This molecule, shown in the centre of the cube in Fig. 18, rotates almost freely since most of the interatomic interactions are cancelled out at this lattice position and only a very weak crystal field remains. Hence, it provides an example of a quasi-free quantum-mechanical rotor whose energy levels are given by:

$$E_J = B\,J(J+1), \qquad J = 0, 1, 2, \ldots, \tag{21}$$

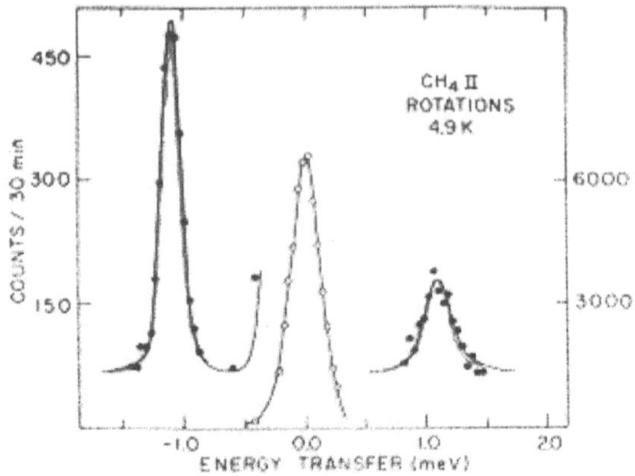

Fig. 20. Energy spectrum of the rotor levels in solid CH_4 (II)

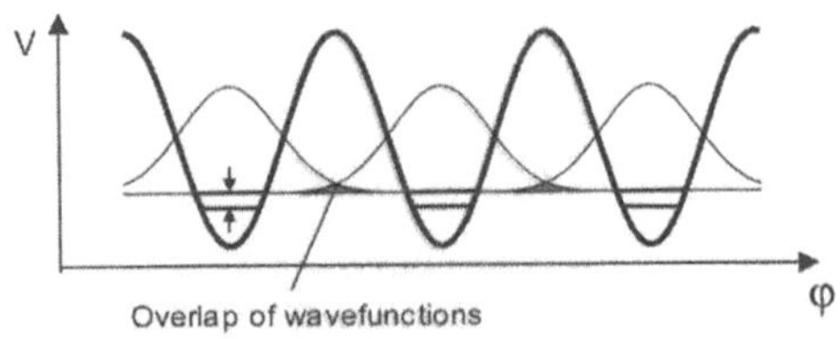

Fig. 21. Threefold rotational potential and the tunnel-splitting of the rotational ground state due to the overlap of wave functions

where B is the rotational constant of the CH_4-molecule.

As demonstrated by *Press and Kollmar* [8], neutrons can induce transitions between the energy levels of the rotator. A corresponding spectrum is shown in Fig. 20. The $J = 0 \leftrightarrow 1$ transition is observed at about $1.2\,meV$ yielding a rotational constant of $0.6\,meV$ in good agreement with the known moment of inertia.

In contrast to the quasi-free molecule, the ordered CH_4-molecules are subjected to a strong angular potential. Due to its symmetry, there are three quantum-mechanical states according to three different wave-functions centered in the individual potential minima as illustrated schematically in Fig. 21. Due to the overlap of these wave-functions there is a splitting of the energy-levels, known as the tunnel-splitting.

The ground-state splitting is extremely sensitive on the potential barrier and, hence, the measurement of the tunnel-levels provide one of the most accurate methods for the determination of interatomic interactions. As a consequence of the site symmetry of a CH_4-group, there are three different tunnel levels A, T, and E corresponding to a total molecular spin of $2\hbar$, $1\hbar$ and 0, respectively. These energy levels can again be probed with neutrons. The energy range is, however, much lower than for the free rotor. Figure 22 shows results of Heidemann et al. [9] using an experimental resolution which is improved with respect to the data of Fig. 20 by more than an order of magnitude.

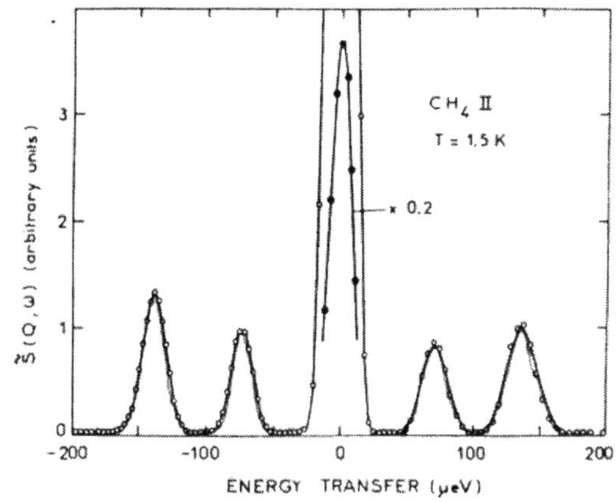

Fig. 22. Tunnel spectrum of CH_4 (II). After [9]

Here, the well defined peaks at $\pm 138\,\mu eV$ and $\pm 73\,\mu eV$ are due to transitions between the tunnel-levels $A \leftrightarrow T$ and $T \leftrightarrow E$, respectively. The third transition $A \leftrightarrow E$ is forbidden due to spin-conservation. This experimental information can be used to determine the angular potential quantitatively.

Similar experiments have been performed in a variety of other systems. In the mixed system $Pb_cSn_{1-c}(CH_3)_4$, the rotation of methyl groups has been observed by Prager et al. [10]. As demonstrated in Fig. 23, the tunnel-spectrum changes drastically with the composition. Obviously, there are at least two components which are due to the rotation of CH_3-groups near tin and lead ions, respectively. The concentration dependence clearly shows that the tunnel splitting is much larger close to Pb-ions and, hence, the potential barrier is significantly lower than in the vicinity of Sn-ions.

4.4 Polymer Dynamics – Between MHz and GHz

The energy range well below $\mu\, eV$ down to $n\, eV$ is accessible by spin-echo spectroscopy. Hence, the dynamics of matter is probed with this technique on a time scale between μs and ns. Spin-echo spectroscopy makes use of the fact that a neutron carries a spin $\hbar/2$ associated with a magnetic moment $-1.913\ \mu_n$ (μ_n = nuclear magneton). Figure 24 illustrates the principle of a spin-echo spectrometer. A rather broad wavelength band of cold neutrons is selected by a velocity selector. This beam is polarised by a magnetic super-mirror and the neutron polarisation is switched to a direction perpendicular to the flight path by a so-called flipper coil. In the precession field $H_\circ$ the neutron spin starts to rotate around the field direction. The precession frequency is given by the Larmor frequency $\omega_L = \gamma H_\circ$, where γ is the gyromagnetic ratio of the neutron. The rotation angle depends on the time the neutron needs to travel through the magnetic field and consequently on its energy. Hence the neutron spin can be regarded as an internal clock.

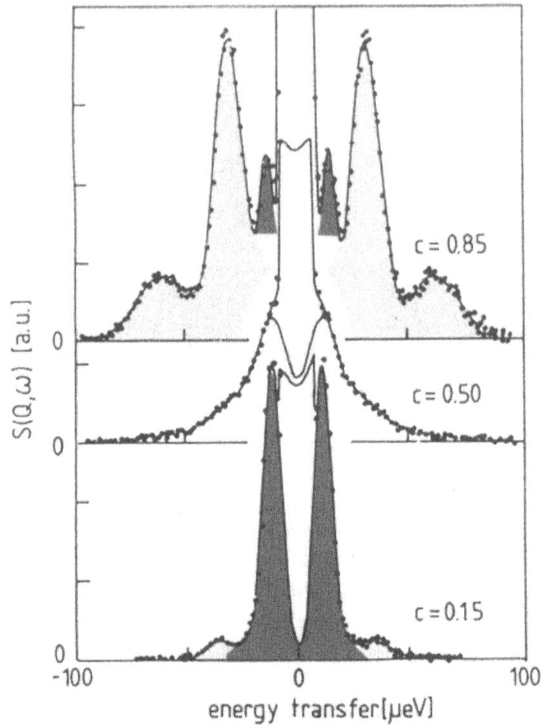

Fig. 23. Tunnel spectra of $Pb_cSn_{1-c}(CH_3)_4$ mixed crystals. After [10]

After the neutron is scattered at the sample the precession is reversed in a second precession field H_1. If there is no energy transfer to the sample and both precession fields are identical the neutron polarisation must be the same as at the entrance of the first coil. These neutrons are able to pass the analyser and are registered in the detector. If, however, the scattered neutrons are slower or faster than the incident ones, the spin-precession is not completely reversed and the neutron polarisation at the analyser is changed unless the field strength H_1 is varied. Interestingly, the depolarisation depends only slightly on the initial neutron energy. Therefore, a broad spectrum can simultaneously be used since each neutron has its own specific clock.

If the intensity is determined as a function of the field strength H_1 a spectrum is obtained which is just proportional to the so-called intermediate scattering function

$$I(\mathbf{Q}, t_F) = \int S(\mathbf{Q}, \omega) e^{i\omega t_F} d\omega = \int G(\mathbf{r}, t_F) e^{i\mathbf{Q}\cdot\mathbf{r}} d\mathbf{r} \tag{22}$$

where the Fourier-time is given by:

$$t_F = \gamma \frac{\hbar H l}{M \bar{v}^3} \tag{23}$$

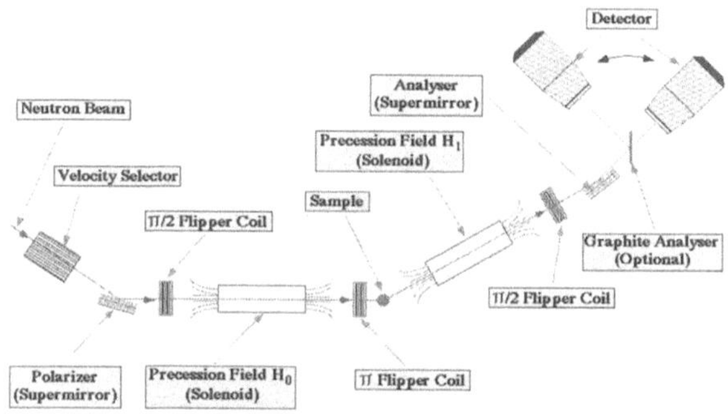

Fig. 24. Layout of the spin-echo spectrometer at the FRJ-2 reactor, Jülich (top) and view of the arrangement of the precession coils (below)

(l is the length of the flight path within the magnetic field H, and $\bar{v}$ is the mean neutron velocity). Hence the experimental data directly yield the frequency Fourier-transform of the scattering functions and, simultaneously, the space Fourier-transform of the van-Hove correlation function. Details of the theory of spin-echo spectroscopy may be found in [11]. It should be noted, that in the last few years new techniques were developed which replace the rather long precession solenoids by pairs of so-called resonance-coils. The discussion of this resonance spin-echo concept is, however, beyond the scope of this lecture. For details, the reader is referred to the literature [12].

Relaxation processes and also slow diffusive motion can be investigated particularly well with spin-echo techniques since correlation times or diffusion coefficients can directly be read from the experimental data. The study of dynamic processes in soft matter like polymers and investigation of changes of conformations is a very active field of current research. A number of models have been put forward to describe the dynamics of polymer systems. Neutron spin-echo spectroscopy was able to check the predictions of those models and to decide which models are appropriate for a quantitative explanation of nature.

In dense polymeric systems like a polyethelene melt, the motion of a single chain molecule is strongly restricted by its surrounding. The reptation model of

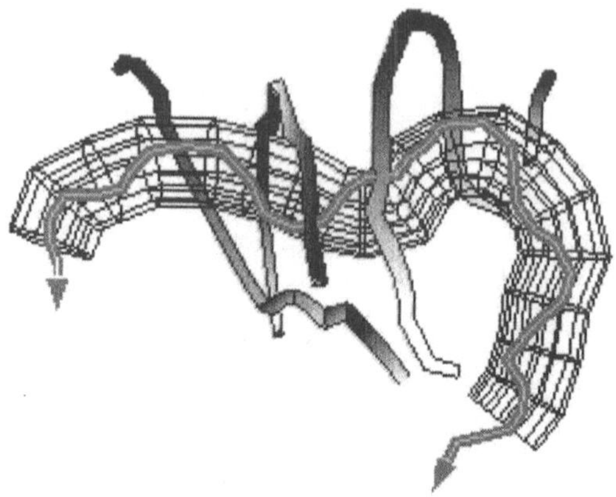

Fig. 25. The reptation concept for highly entangled polymer systems. The motion of the tagged polymer molecule is restricted to a tube defined by the surrounding molecules . After [14]

de Gennes [13] assumes that each individual polymer molecule is kept within a tube defined by the surrounding molecules. The motion perpendicular to the tube axis is strongly restricted and relaxational processes are more or less confined to the direction along the tube. Hence, the polymer chain moves like a snake along its own contour line as illustrated in Fig. 25, a behaviour which led to the term "reptation". The longitudinal relaxation time is expected to be much longer than transverse ones which should lead to characteristic features in the intermediate scattering function.

Indeed the spin-echo study of *Schleger et al.* [14] on highly entangled polyethelene (PEB-2) proves that the experimental data are consistent with the predictions of the reptation model while other models show significant discrepancies.

In Fig. 26 some results are presented. The normalised intensity (denoted by $S(\mathbf{Q},\omega)/S(\mathbf{Q},0)$ is plotted versus the time t for two different values of the scattering vector. The sharp decrease is due to the transverse fluctuations while the extended plateau is characteristic for the much slower longitudinal relaxation.

Obviously, spin-echo spectroscopy allows to resolve dynamical processes on time scales up to several 100 ns. Hence, the gap between neutron scattering (fast processes) and NMR-investigations (in the MHz regime) becomes smaller and smaller.

4.5 Kinetic Studies – Real-Time Experiments

Usually, inelastic neutron-scattering experiments require rather long counting times. This is due to the limited source flux provided by the existing research reactors and spallation source, on the one hand, and to the rather small scattering cross sections of most substances, on the other hand. As a rule of thumb, phonon intensities, e.g., are about three or four orders of magnitude weaker than structural Bragg reflections. While modern focussing techniques, advanced neu-

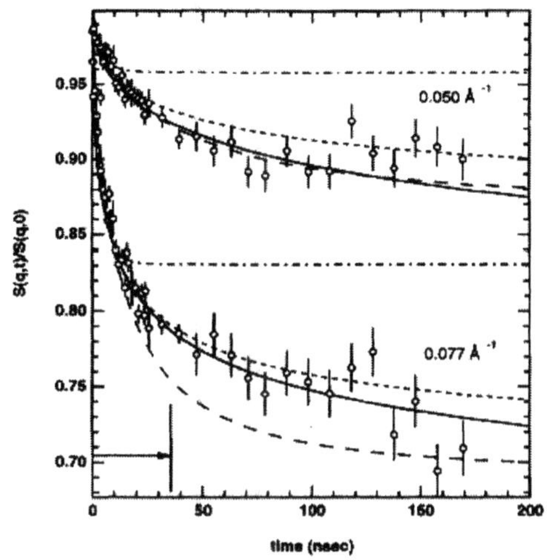

Fig. 26. Spin-echo data for a highly entangled poly-ethelen melt (PEB-2) at two different scattering vectors. The lines are the results of model calculations. The best agreement is found for a reptation model. After [14]

tron guides, multi-detectors e.g., help a lot to improve the beam utilisation, the counting times are nevertheless of the order minutes rather than milliseconds. Therefore, kinetic studies on a second or subsecond time-scale were beyond the scope of neutron scattering. If the processes under consideration are, however, reversible or at least repeatable, stroboscopic techniques can be used to explore the time evolution of chemical processes or phase transitions even with a time-resolution of microseconds [15].

Recently, the evolution of phonon spectra during the decomposition in mixed ionic crystals have been observed by time-resolved three-axes spectroscopy [16]. As a model system AgBr-NaBr was used which exhibits a simple phase diagram with a well defined almost symmetrical miscibility gap and an upper critical point of about 558 K (cf. Fig. 27). Since the pure constituents show rather different elastic constants, the phase separation can be monitored on a microscopic scale by the time-variation of the acoustic phonon spectra.

Mixed single-crystals were periodically heated up to temperatures of 630 K to allow homogenisation and then quenched into the miscibility gap in order to observe the kinetics of demixing. For each setting $(\mathbf{Q}, \omega)$ of the spectrometer, neutrons were counted as a function of time and the intensities were accumulated over a number of temperature cycles. Figure 28 shows the spectrum of a transverse acoustic phonon as a function of time after the quench deep into the miscibility gap. At the beginning, a single phonon peak is observed, which corresponds to the homogeneous phase. After about 200 s, the shape of the phonon peak has changed drastically due to the beginning decomposition. At 345 s, one can clearly distinguish two phonons which correspond to the two product phases. Hence, the phase separation takes place on a rather short time scale.

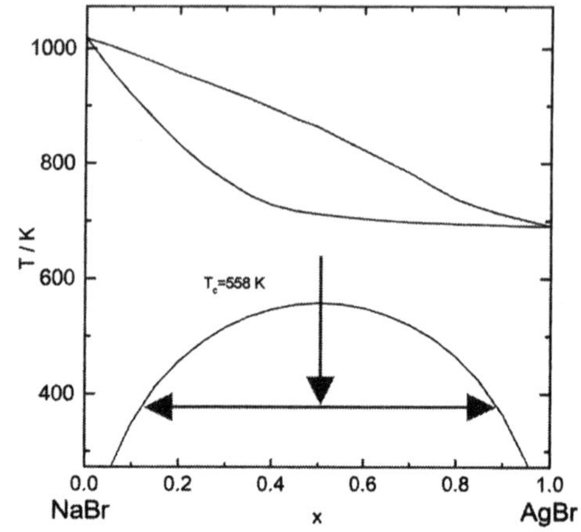

Fig. 27. Miscibility gap of the system AgBr-NaBr

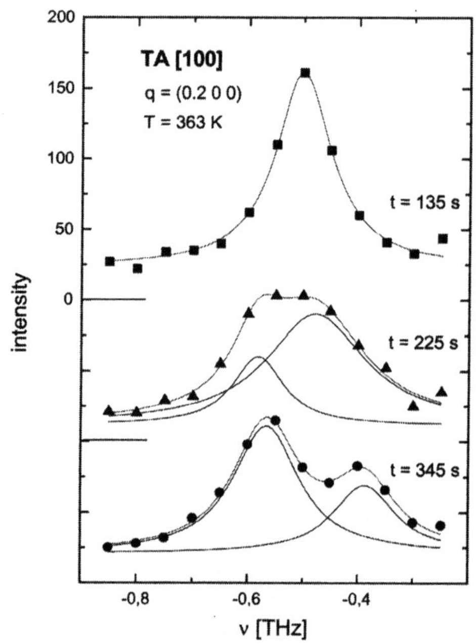

Fig. 28. Time evolution of transverse acoustic phonon spectra of $Ag_{0.35}Na_{0.65}Br$ after quench to 363 K. After [16]

The corresponding time dependence of the phonon frequencies is shown in Fig. 29. Interestingly, during this time-interval, the Bragg reflections remain almost unchanged, even though the pure constituents exhibit different lattice parameters. Obviously, there are strong coherency strains which keep the lattice in a metastable state. The chemical demixing as reflected by the variation of interatomic forces and, hence, by the phonon frequencies, on the one hand,

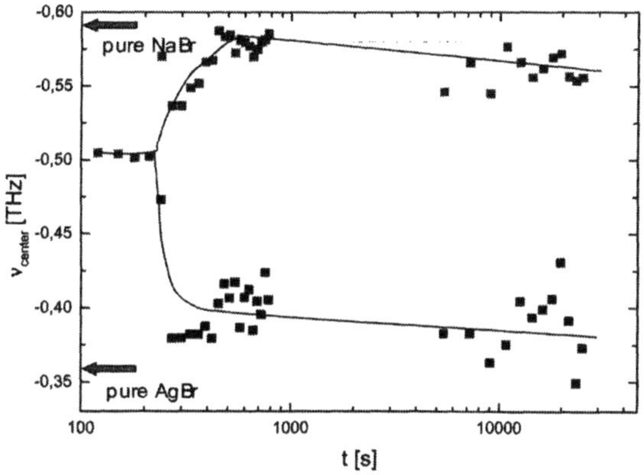

Fig. 29. Time dependence of the phonon frequency in $Ag_{0.35}Na_{0.65}Br$ after a quench to 363 K. After [16]

and the mechanical relaxation of the underlying lattice, on the other hand, are different aspects of the decomposition process.

Time-resolved phonon investigations thus provide the most direct information about the microscopic changes in solids.

5 Conclusions

In this contribution, some applications of inelastic neutron scattering have been presented which seem to be representative for the investigation of dynamical properties of matter. Of course, there is a wealth of different interesting topics which could not be mentioned within a single lecture, but I hope that the reader received an impression about the power of this experimental method.

I want to stress again, that there is another extremely interesting field of applications which was beyond the scope of the present paper and had to be skipped completely - it is the study of magnetic excitations. The variety of different magnetic phenomena which may be probed with neutrons would, however, require another review.

References

1. e.g. Vorlesungsmanuskripte, 27. IFF-Ferienkurs "Streumethoden zur Untersuchung kondensierter Materie", Forschungszentrum Jülich 1996, ISBN 3-89336-180-4 Lecture Notes, Summer School on Neutron Scattering 1993, PSI-Proceedings 93-01, ISSN 1019-6447.
2. e.g. S.W. Lovesey, Theory of Neutron Scattering From Condensed Matter, Clarendon Press, Oxford 1984 , ISBN 0-19-852015-8.
3. L. Pintschovius, N. Pyka, W. Reichardt, A.Y. Rumiantsev, N.L. Mitrofanov, A.S. Ivanov, G. Collin, P. Bourges, Physica C **185-189**, 156 (1991).

4. J.C. Marmeggi, R. Currat, A. Bouvet, G.H. Lander, Physica B **263-264**, 624 (1999) and ILL Annual Report 98, p. 16.
5. S. Klotz, J.M. Besson, M. Braden, J. Kulda, Annual Report, ILL 1998, p. 44.
6. J. Texeira, M.C. Bellisent-Funel, S.H. Chen, A.J. Dianoux, Phys.Rev. A **31**, 1913 (1985).
7. H. Jobic, M. Bée, Annual Report, ILL 1997, p. 16.
8. W. Press, A. Kollmar, Solid State Commun. **17**, 405 (1975).
9. A. Heidemann, W. Press, K.J. Lushington, J.A. Morrison, J. Chem.Phys. **75**, 4003 (1981).
10. M. Prager, Da Zhang, A. Weiss, Physica B **180-181**, 671 (1992).
11. F. Mezei (ed.), "Neutron Spin Echo", Lecture Notes in Physics, Springer, Berlin 1980.
12. e.g. M. Köppe, M. Bleuel, R. Gähler, R. Golub, P. Hank, T. Keller, S. Longeville, U. Rauch, J. Wuttke, Physica B **266**, 75 (1999).
13. P.G. de Gennes, J. Physique **42**, 735 (1981).
14. P. Schleger, B. Farago, C. Lartigue, A. Kollmar, D. Richter, Phys. Rev.Lett. **81**, 124 (1998); Annual Report, ILL 1998, p. 68.
15. G. Eckold, Nucl.Instr.& Methods A **289**, 221 (1990).
16. P. Elter, G. Eckold, D. Caspary, F. Güthoff, A. Hoser, Appl. Phys. A, in press.